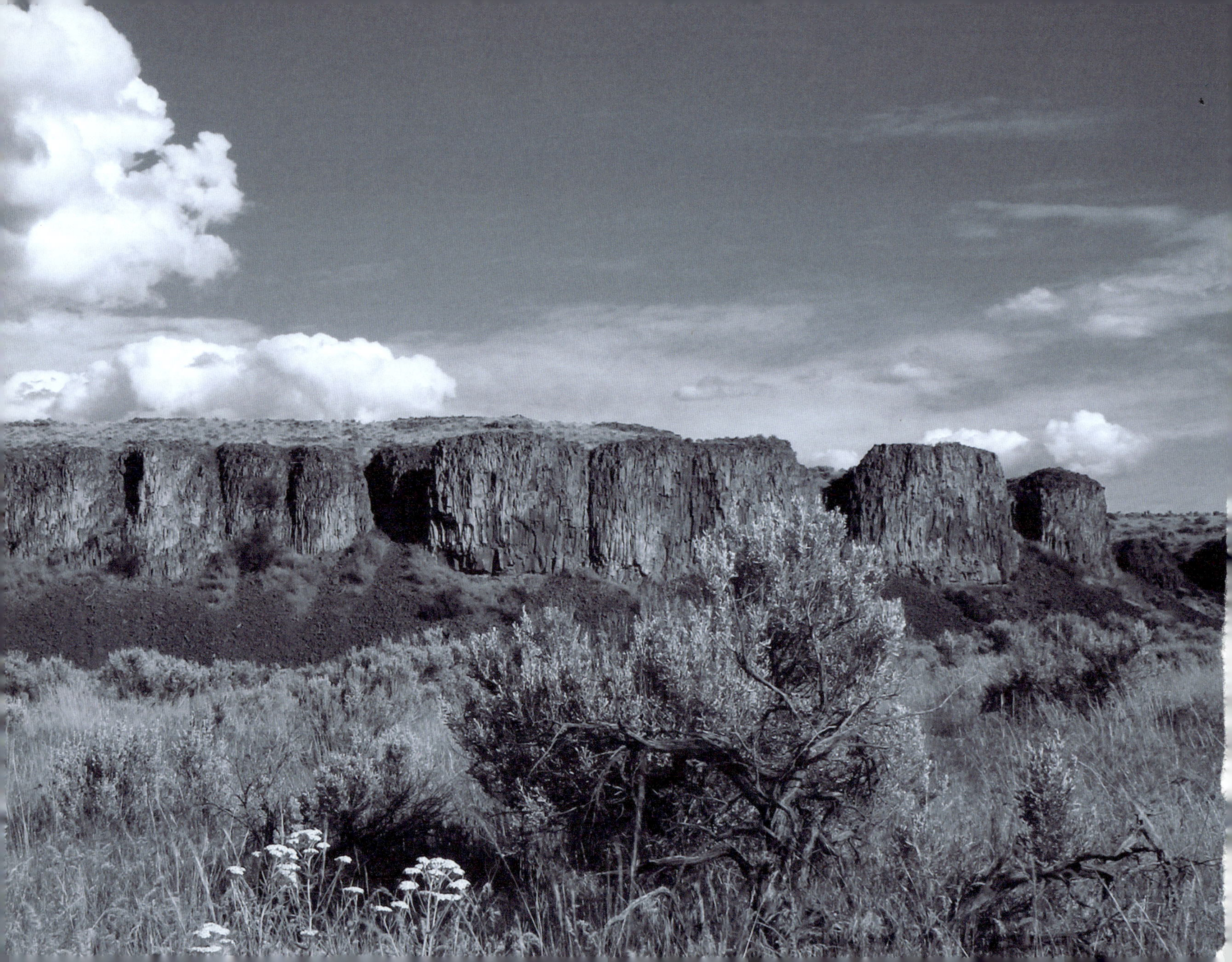

BEAUTIFUL WOUNDS

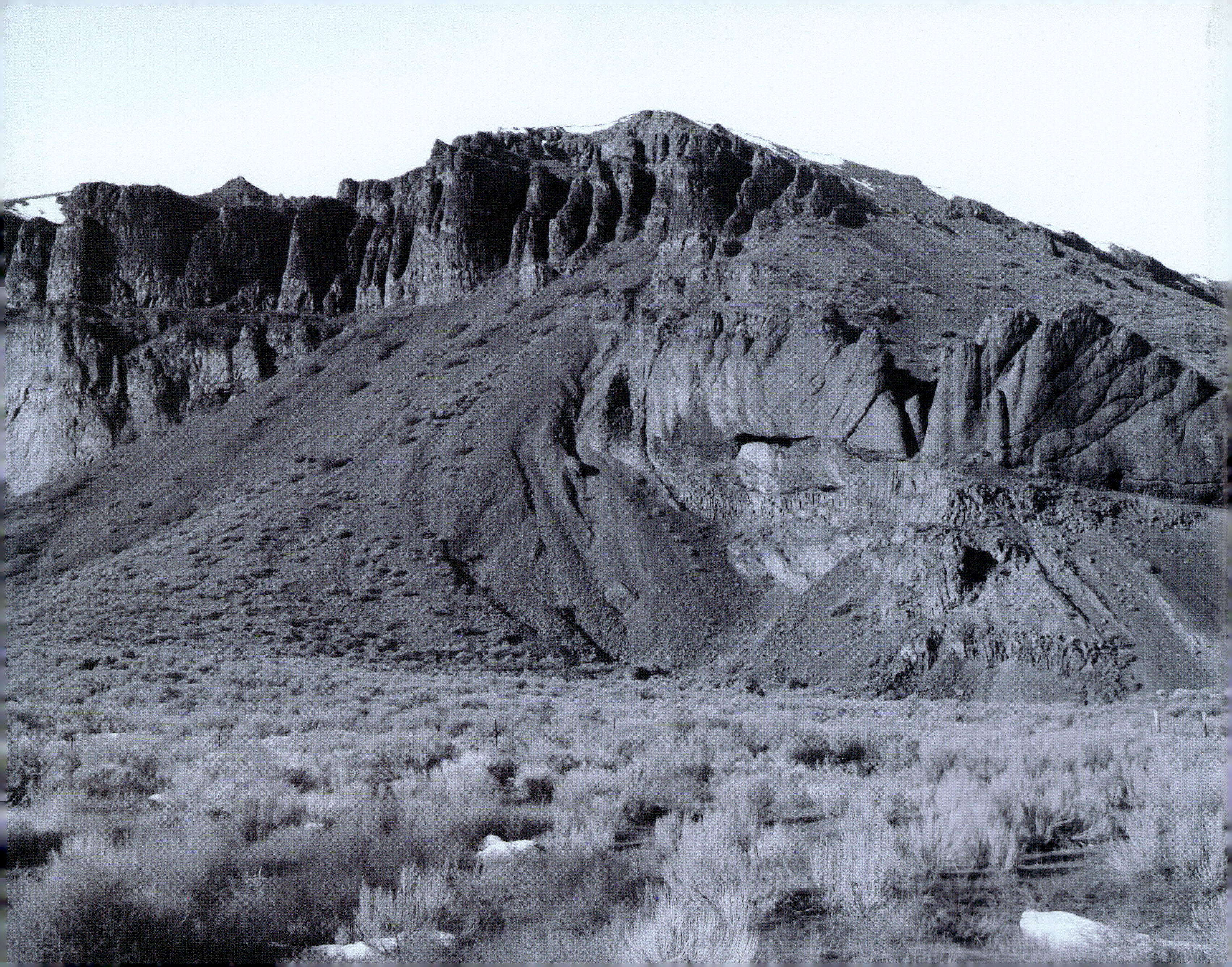

BEAUTIFUL WOUNDS

A Search for Solace and Light in Washington's Channeled Scablands

TIMOTHY CONNOR

The Countryman Press

An Imprint of W. W. Norton & Company
Independent Publishers Since 1923

Printed in China

Page 34: Photo of J Harlen Bretz courtesy of the Hanna Holborn Gray Special Collections Research Center, University of Chicago Library

Manufacturing by Toppan Leefung Pte. Ltd.
Book design by Allison Chi
Production manager: Devon Zahn

The Countryman Press
www.countrymanpress.com

A division of W. W. Norton & Company, Inc.
500 Fifth Avenue, New York, NY 10110
www.wwnorton.com

978-1-68268-680-5

10 9 8 7 6 5 4 3 2 1

Endpapers: (map) Map of *The Channeled Scablands of the Columbia Plateau and Their Associated Features by J Harlen* Bretz, 1923; (front) East wall, Dry Coulee, Grant County; (back) Southern rim, Wallula Gap, Benton County.

Previous: Winter range, lower Grand Coulee.

For our beloved Audrey, without whom I simply wouldn't be.

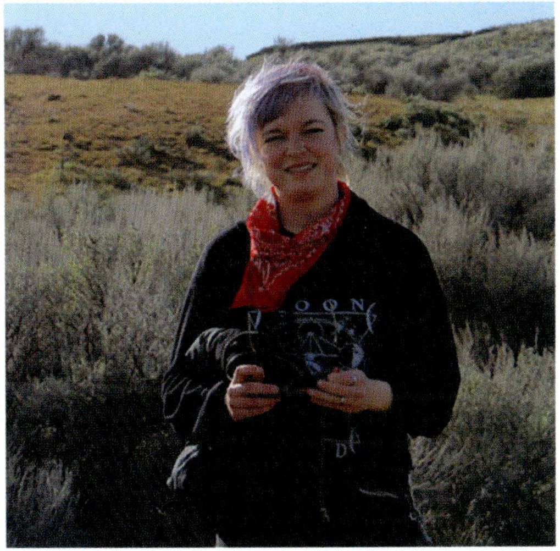

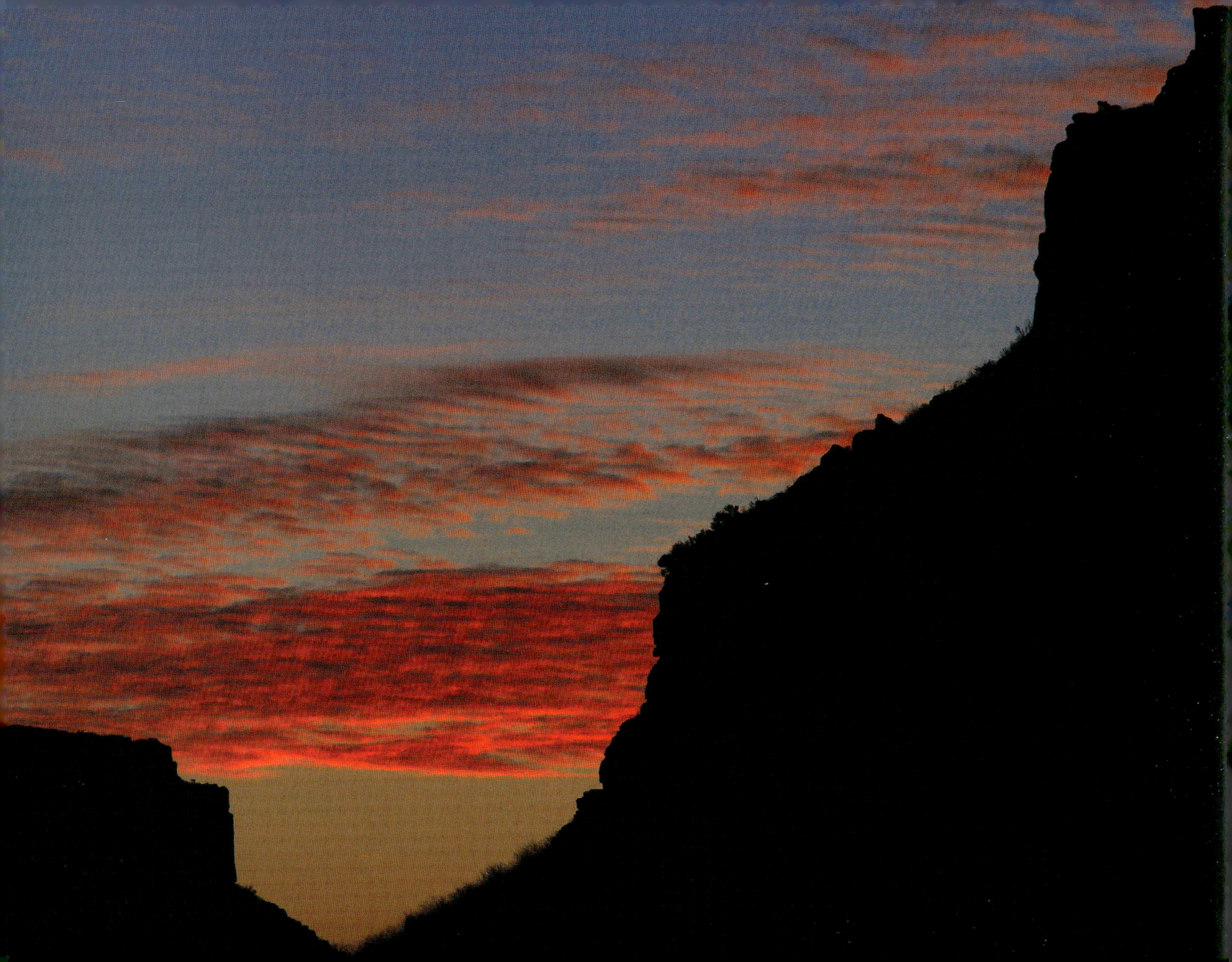

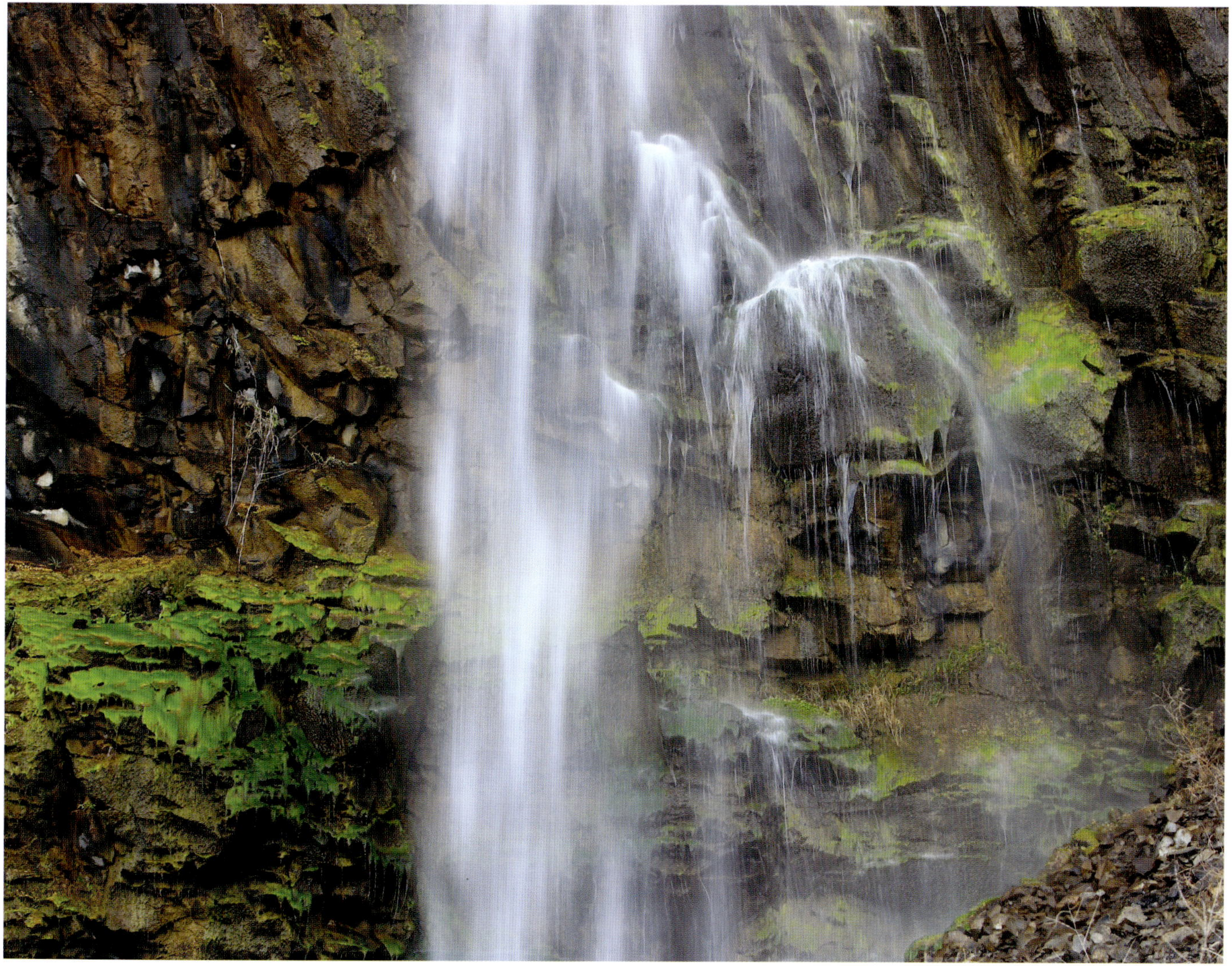

Left: Canyon sunrise, Drumheller Channels. *Right:* At the foot of the falls, Frenchman Coulee.

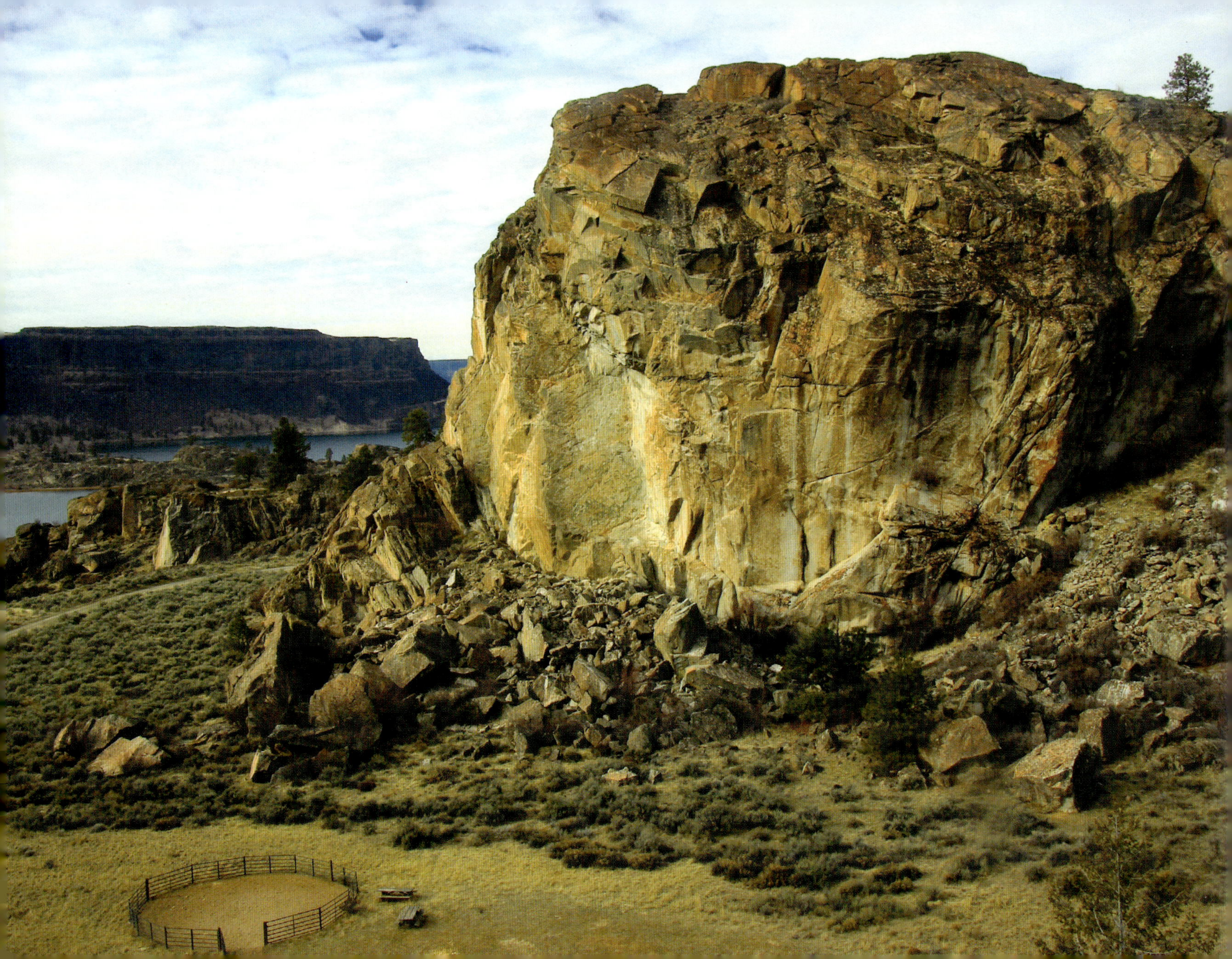

CONTENTS

Left: Equestrian area, Northrup Canyon, upper Grand Coulee.

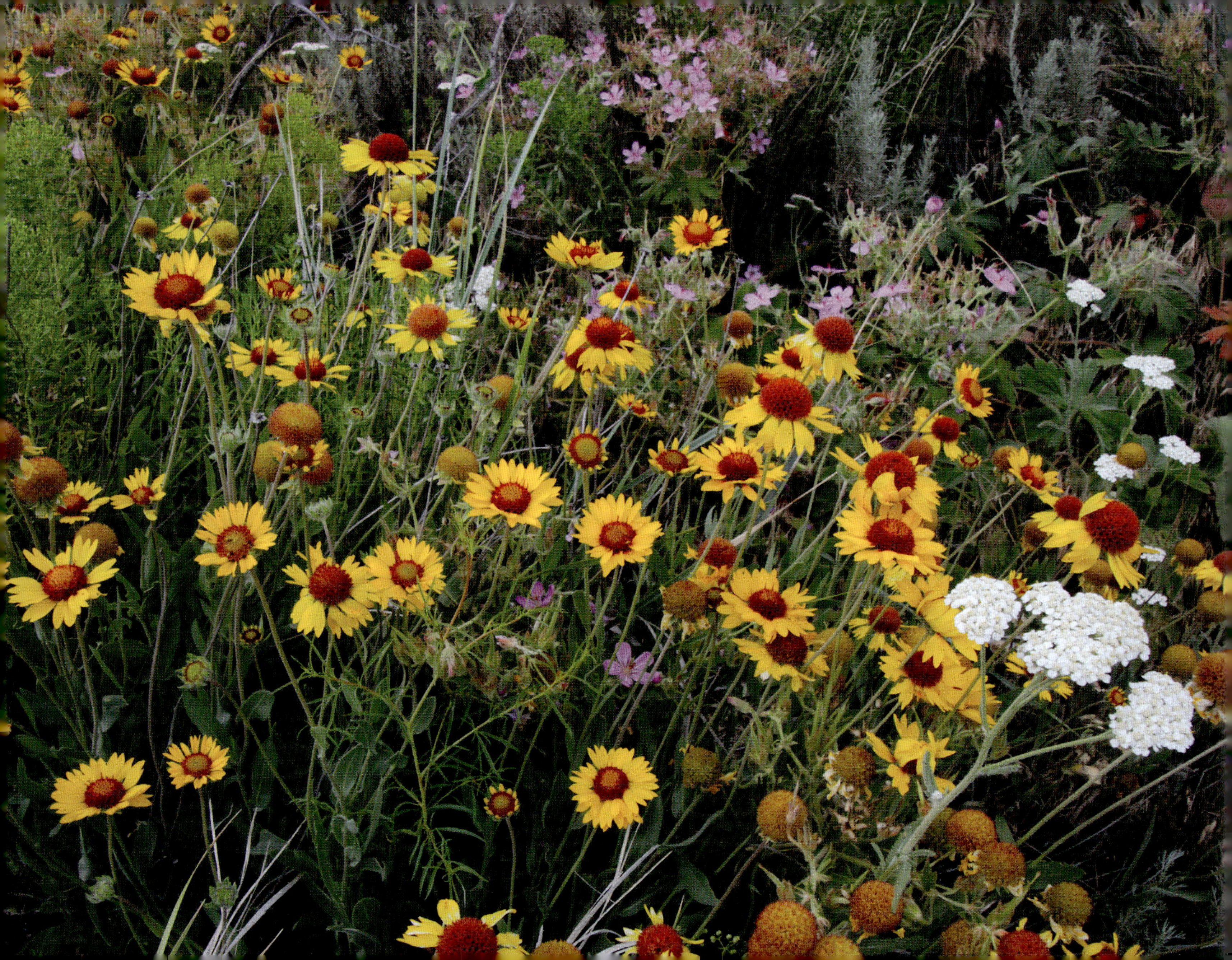

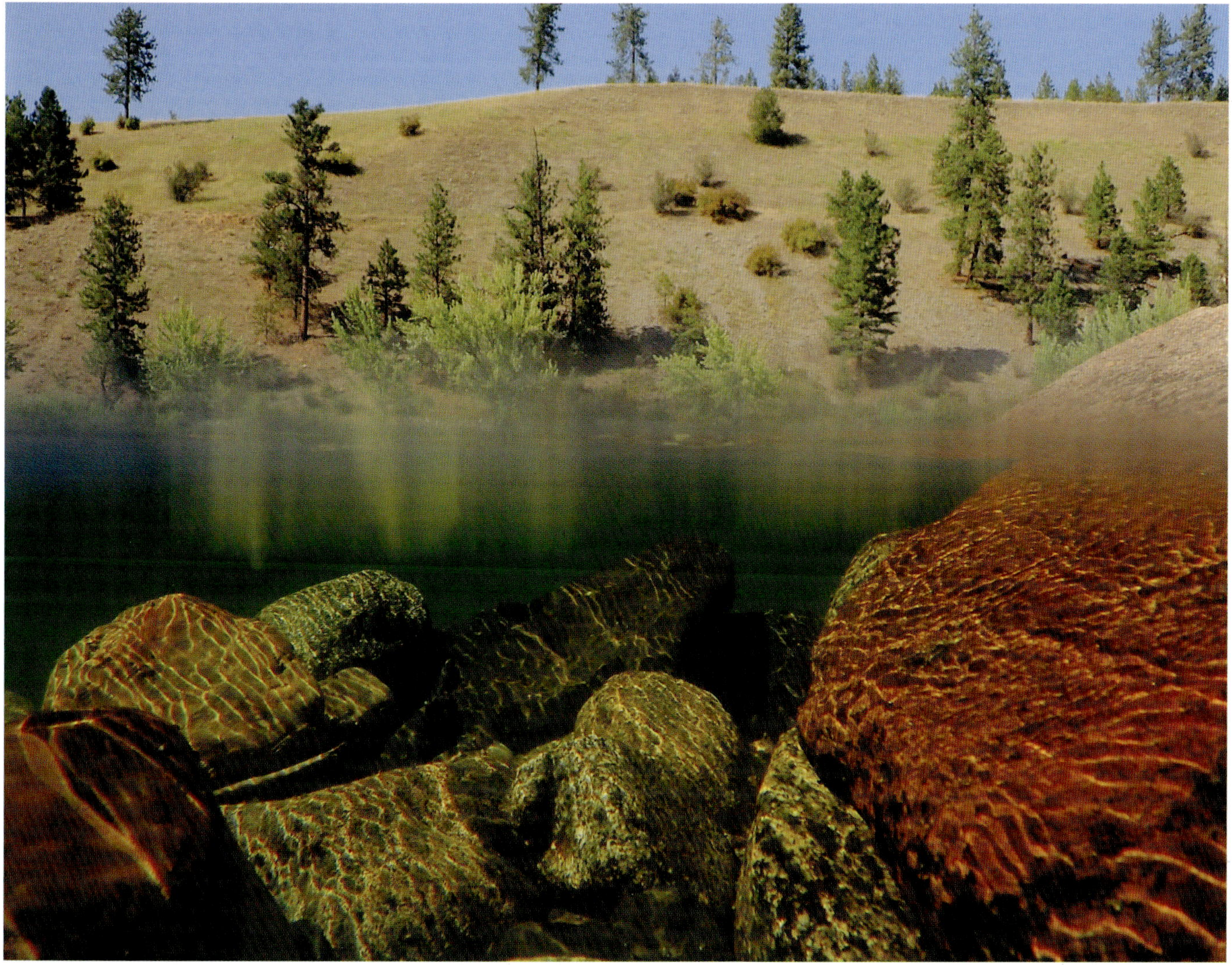

Left: Scabland bouquet, Adams County. *Right:* Ice age flood deposit, Spokane River, West Spokane.

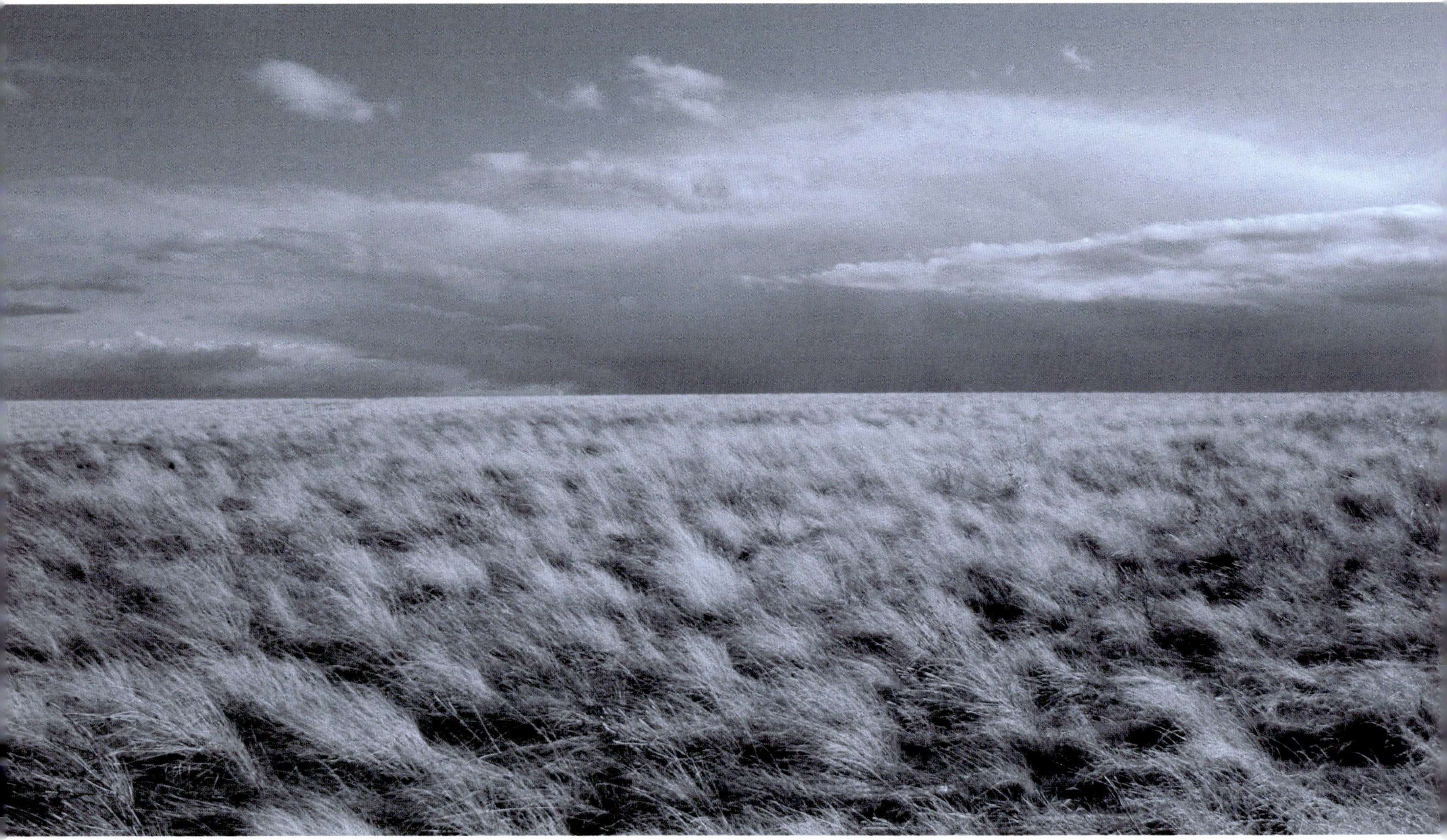

Left: Grasslands above the Snake River, Franklin County. *Right:* American kestrel, Whitman County.

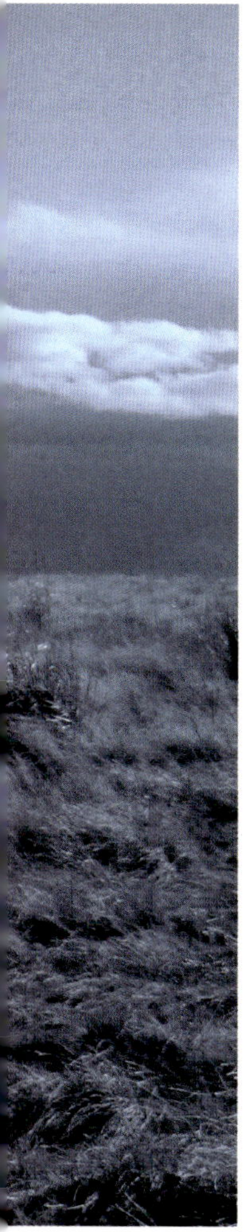

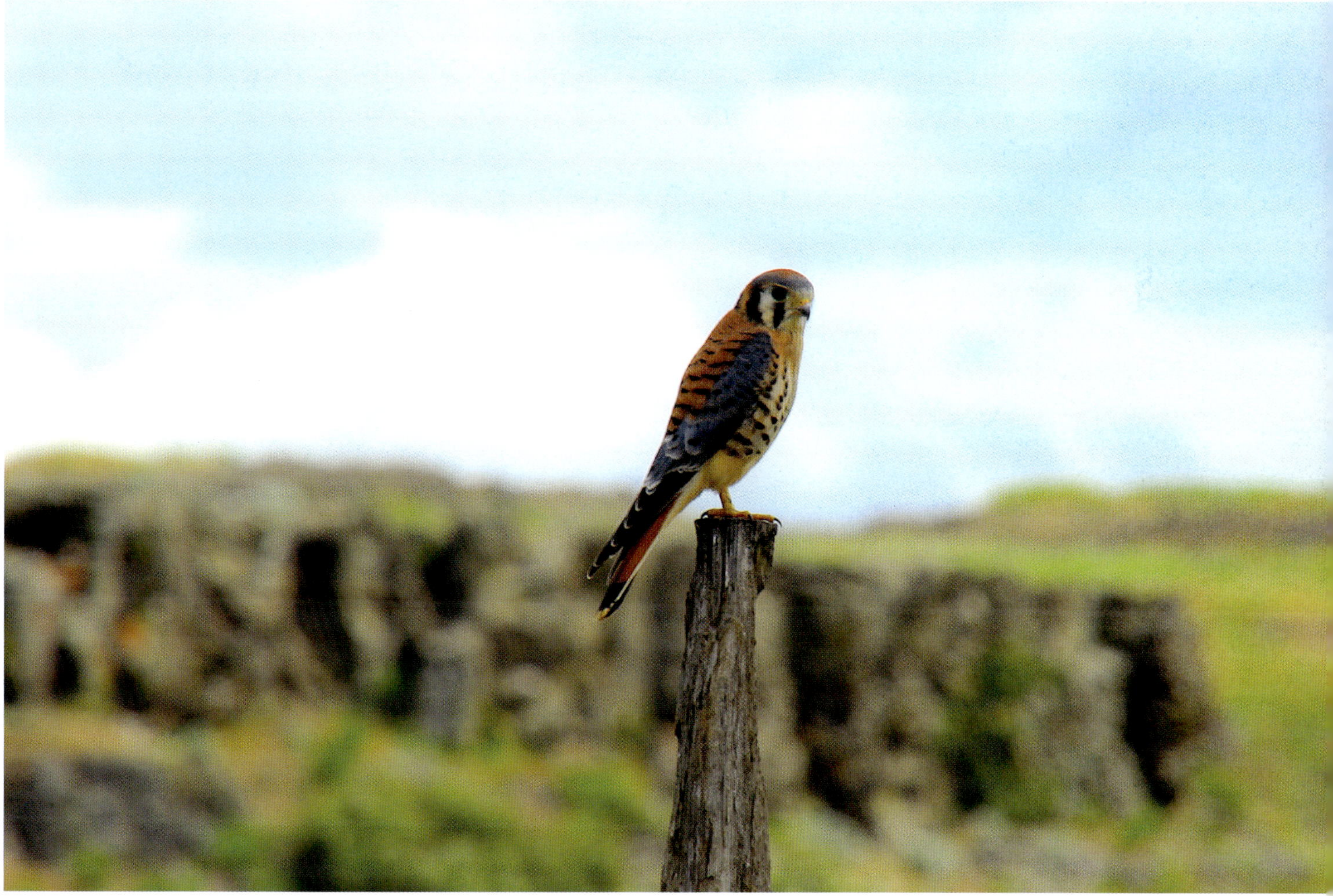

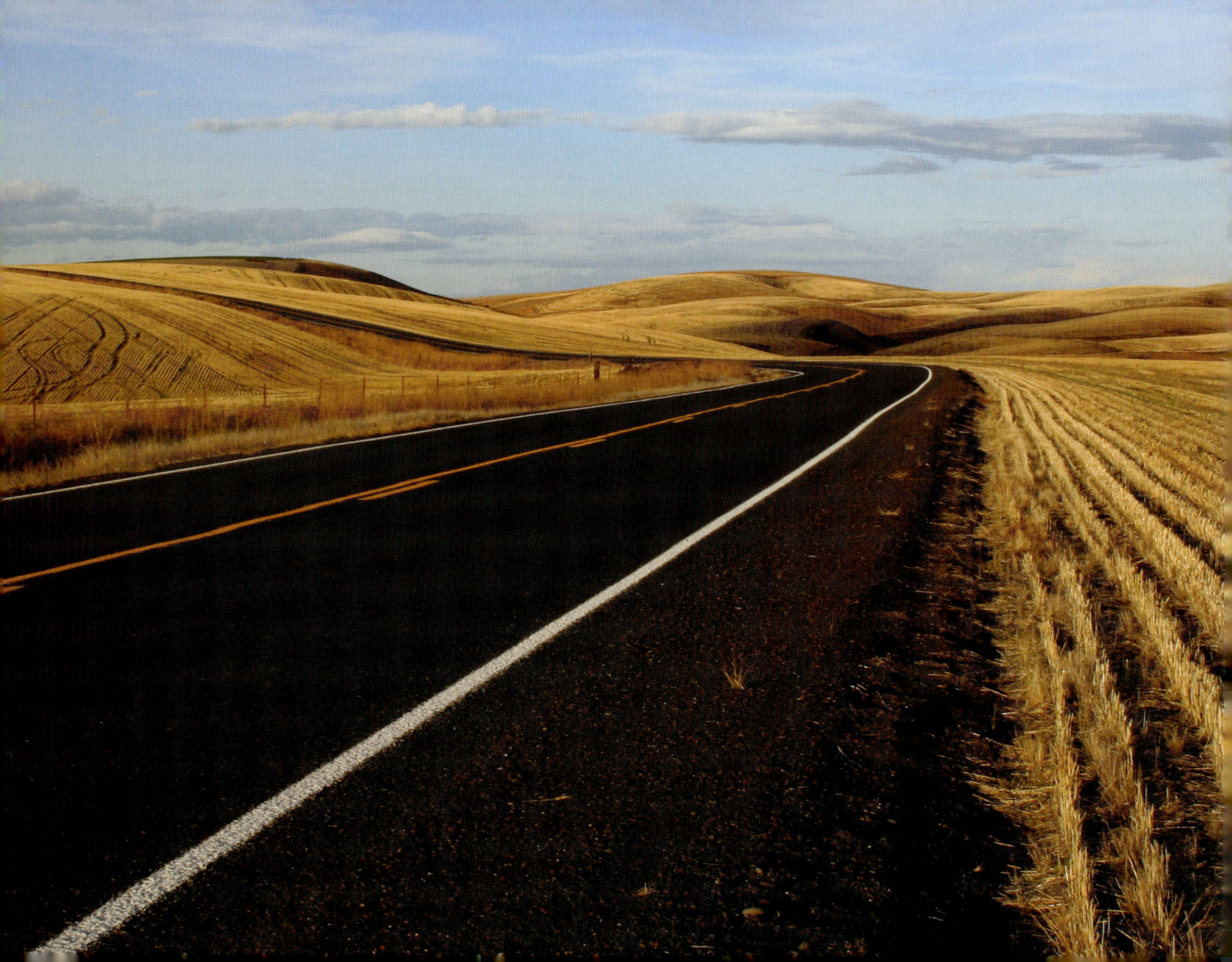

PROLOGUE

The Road to Kahlotus

The path toward this book began in the lingering light of a late afternoon. I was leaving Pasco, a town near the confluence of the Snake and Columbia Rivers, where my mother had been born and raised. Much of me is from Pasco as well. It is where I was conceived. It is where the scent of sagebrush and the aroma of my sweet grandmother's biscuits and fried chicken had been applied, like a base coat of lacquer, onto the emerging shelves of my memory.

A drive from Pasco to my home in Spokane can take as little as two hours. But on that day, a decade ago, I didn't want to be home in two hours. Instead, I chose back roads that would take me into the channeled scablands, looping to the northeast, generally toward Spokane, but first into the rugged and sparsely populated plateau above the Snake.

Seventeen miles east of Pasco, Pasco Kahlotus Road leaves a nest of irrigation circles and starts a climb into bare hills near the Juniper Dunes Wilderness. It was at this juncture where something inside me snapped. I began trembling and sobbing. In retrospect, I was being arrested by a jackbooted truth. I pulled off the road to try to gather myself.

My wife and I had been together twenty years and were raising two wonderful children. We were less alike than we'd hoped and, in the past few years, a cascade of tragic events had finally overwhelmed us, eaten away at our finances, and broken our bond. The denouement came not long after my trip home. I woke at dawn to notice a frown on her face. I nudged her awake to ask with a whisper if she was okay. And then she would be the one who was

Left, Pasco-Kahlotus Road near Juniper Dunes.

trembling, starting the day by telling me she wanted a divorce. It would be amicable, which is not to say it wouldn't also be excruciating.

In what seemed at the time like a fortuitously short interval, I fell in love again. It ended with a formal and cordial appointment at which the new love of my life looked me bravely in the eyes and told me she would be marrying someone else. This was amicable as well. On the pain meter, amicable heartbreak is not so different from the other kind.

In the dark months that followed, I realized I'd had choices the afternoon I wound up on the road to Kahlotus. If it was just a matter of killing time before showing up at home at the end of the day, there were plenty of bars to choose from. And yet I'd headed out into the desert beneath a gathering sky. I must have known something dreadful was stalking me. I must have known that a road less traveled might surface a hard truth. I must have wanted to know what I didn't want to admit.

One of the harder truths is that I was mortified by the prospect of being alone, of living without a romantic partner. It was not an epiphany from therapy. It was more like a punch to the solar plexus. It was also a signal that I needed to work on myself so that I could at least enjoy the wonders of life without needing someone to share and verify my experiences. Toward that purpose I resolved to walk alone, at least until I could do so without feeling lonely, discarded, or irrelevant. And this is what drew me into the high desert and the canyons.

The solace of the landscape—this landscape—comes in form and through memory. The scablands got their name because, at least in the eyes of soil-hungry European pioneers, the earth looked badly wounded. The bones of the bedrock are not just exposed; they're often toppled and broken. In dry outwash basins they litter the terrain as though they're leftovers from a colossal meal wolfed down by mythical predators.

But there is also ethereal beauty. Without much soil to soak up snowmelt and vernal rain, wetlands form. Mule deer graze and wildflowers bloom. Larks and canyon wrens sing their songs. A rock-rimmed crease in the earth can become a refuge where a heart can warm and heal.

As for memory, I learned the legend of J Harlen Bretz in 1977, when I was studying geology as an undergraduate at Washington State University in Pullman. From Pasco, the road to Kahlotus is not just the road to Pullman; it is also a route into the heart of the landscape that Bretz strode into a century ago to solve the puzzle of the mysteriously large and interconnected scars. And solve it he would, with brilliant science and elegant writing. It's just that his most prominent peers dismissed his work and scoffed at his theory of a great ice age flood.

Perhaps the most moving part of John Soennichsen's 2008 biography, *Bretz's Flood*, is the memory Bretz's daughter shared about how shaken and depressed her father was in the years after his work had been so rudely rejected. And yet this sad part of the Bretz legend also brings to mind the visible glee with which my geology professor and his teaching assistants (it was a large class) touted Bretz's long-overdue vindication. In 1979, at the age of 96, the venerable geologist was finally decorated with the Penrose Medal, the most prestigious award given to American geologists. He died two years later.

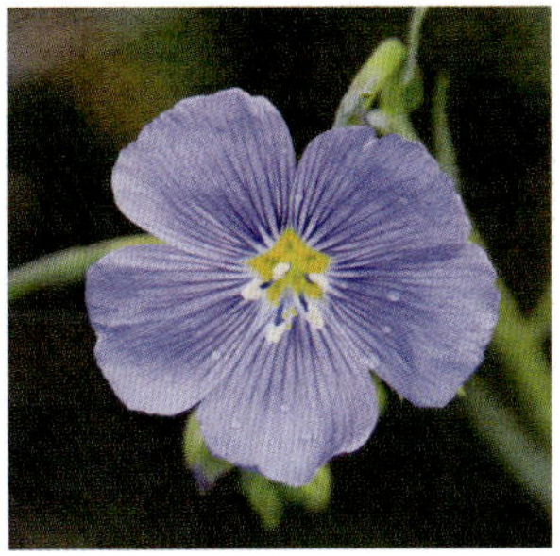

Walking into the scablands was the best therapy I could have prescribed for myself. I've made bloody holes in myself crossing barbed wire, jumped over rattlesnakes, and been forced to dig my car out from snowdrifts. And it's all been to the good. With scant exception, the people who live and work in these remote places have been remarkably kind and helpful. They'll often stop to ask if I'm okay. I smile and try to give them the short answer. This book is the long one.

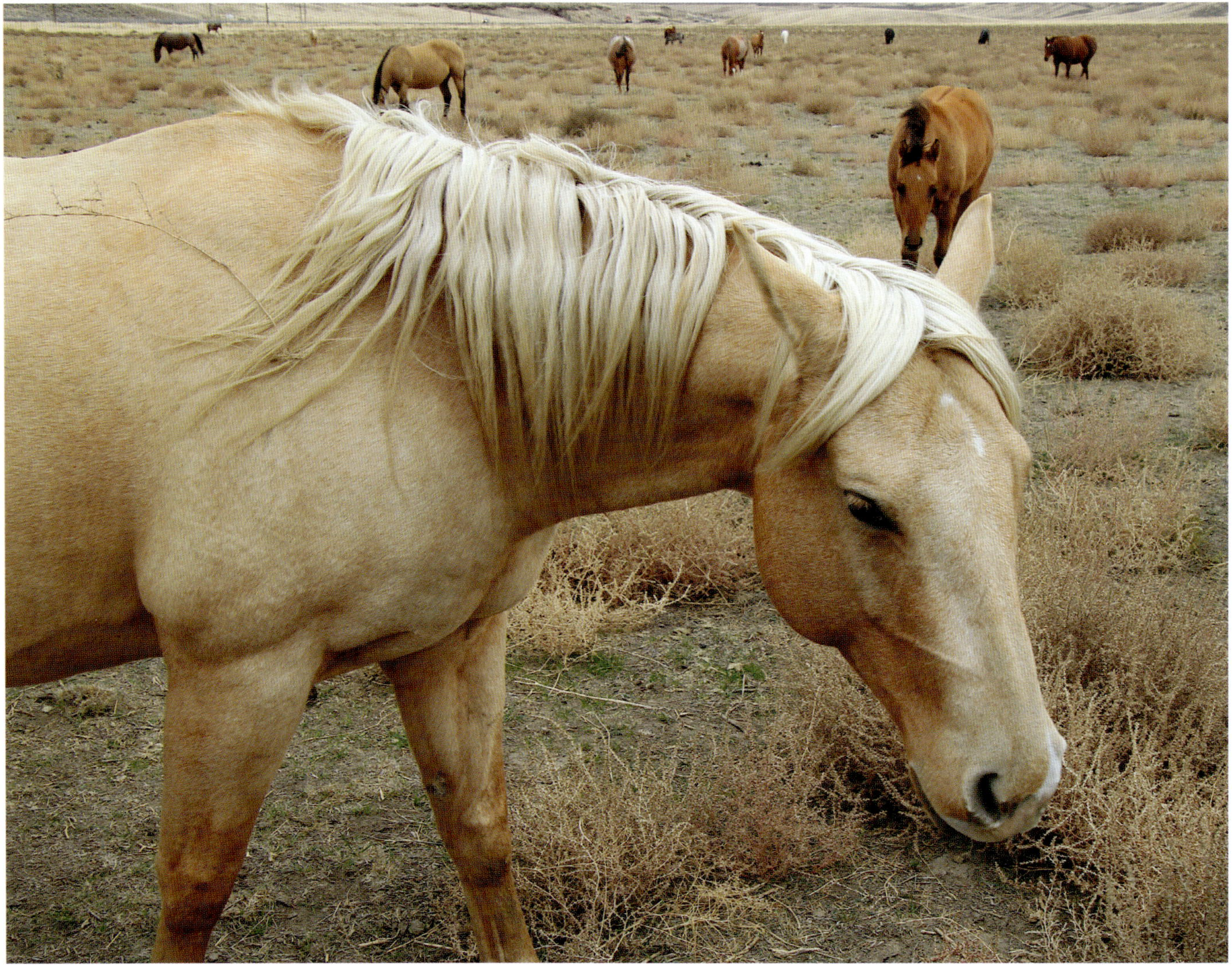

Left: Horses at Washtucna Coulee, Franklin County. *Right:* Frozen Lake, Lake Creek Coulee, Lincoln County.

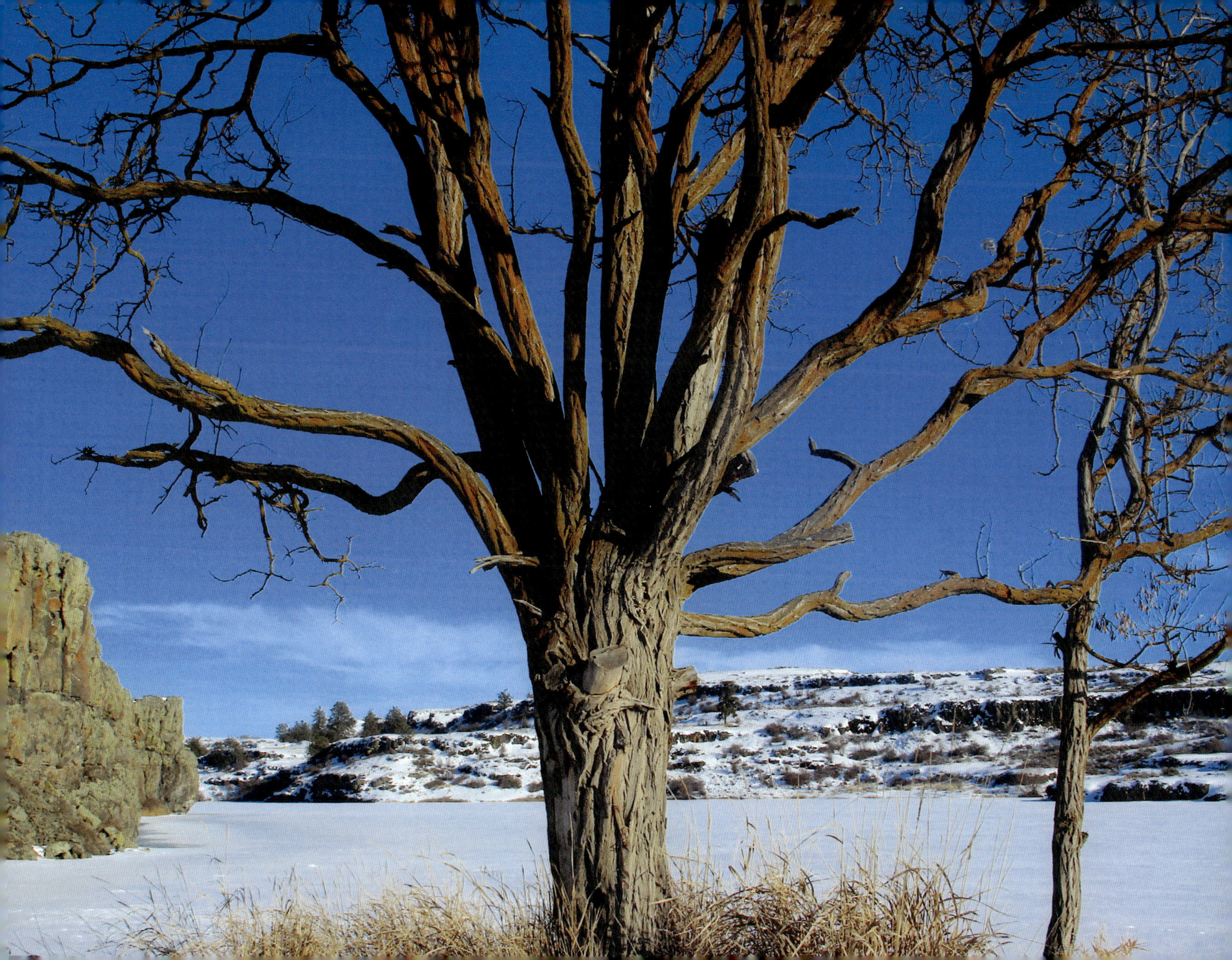

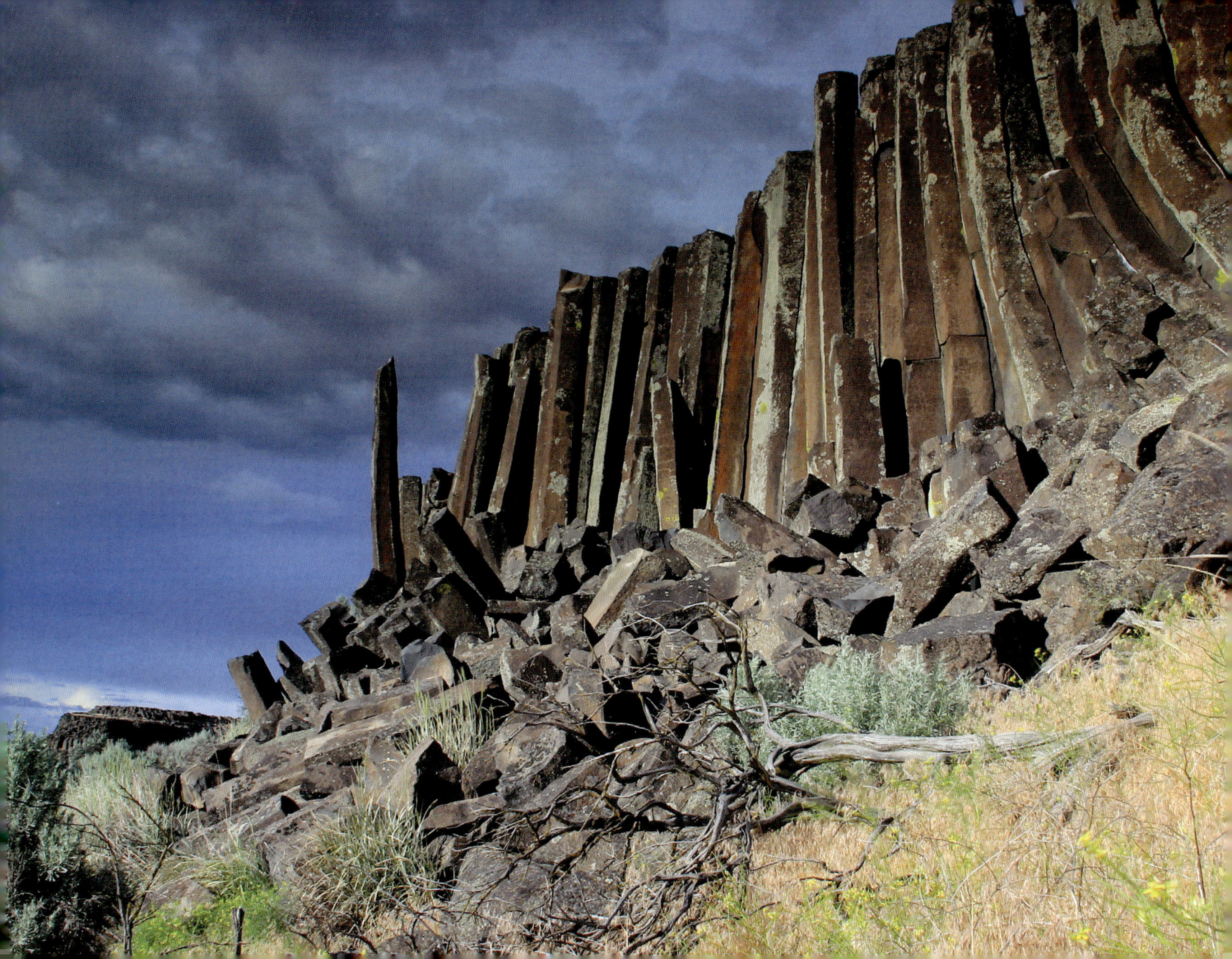

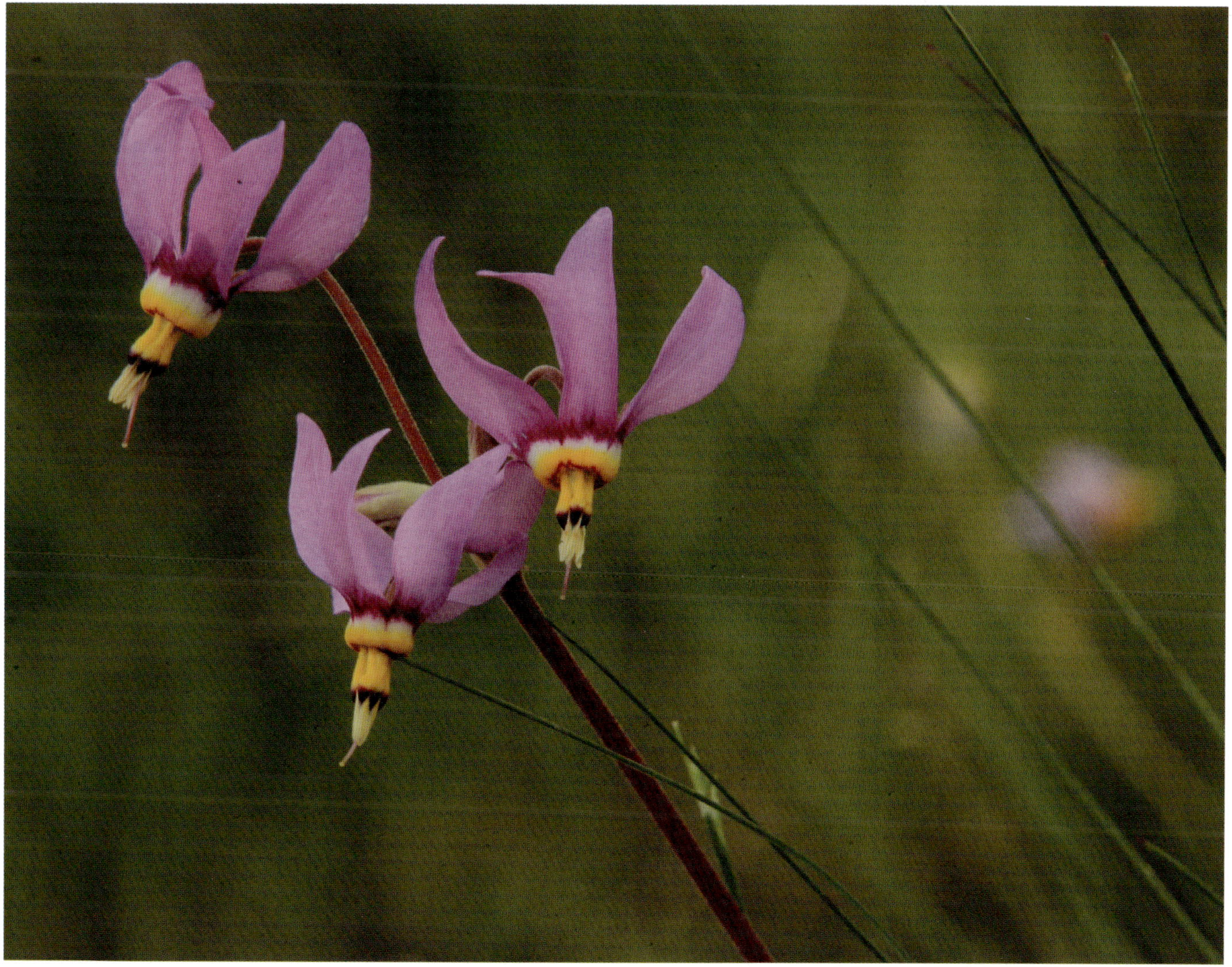

Left: Elephant Mountain basalt palisade, Drumheller Channels. *Right:* Bonneville shooting star.

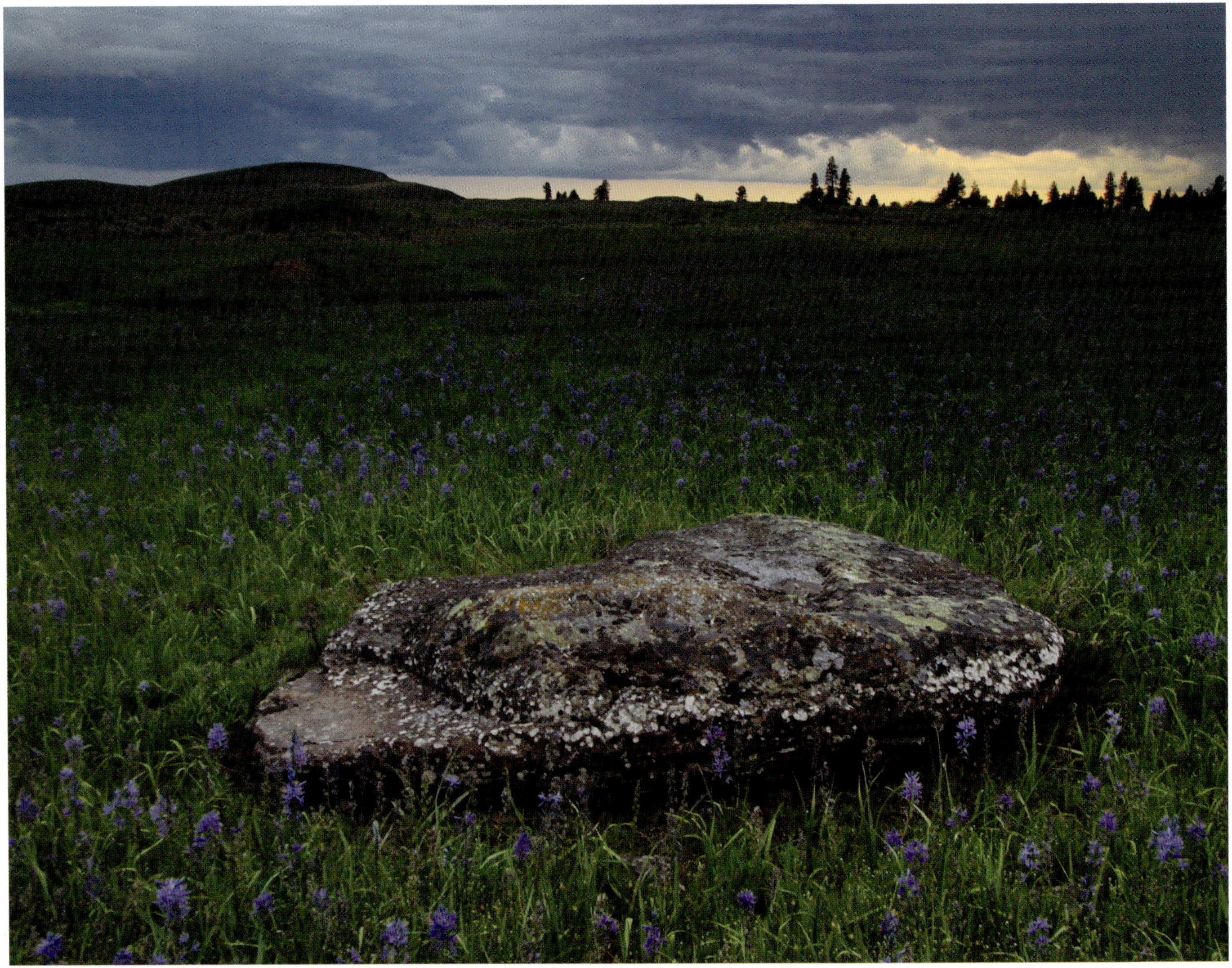

Left: Storm on Miller Ranch, Lincoln County. *Right:* Carey's balsamroot.

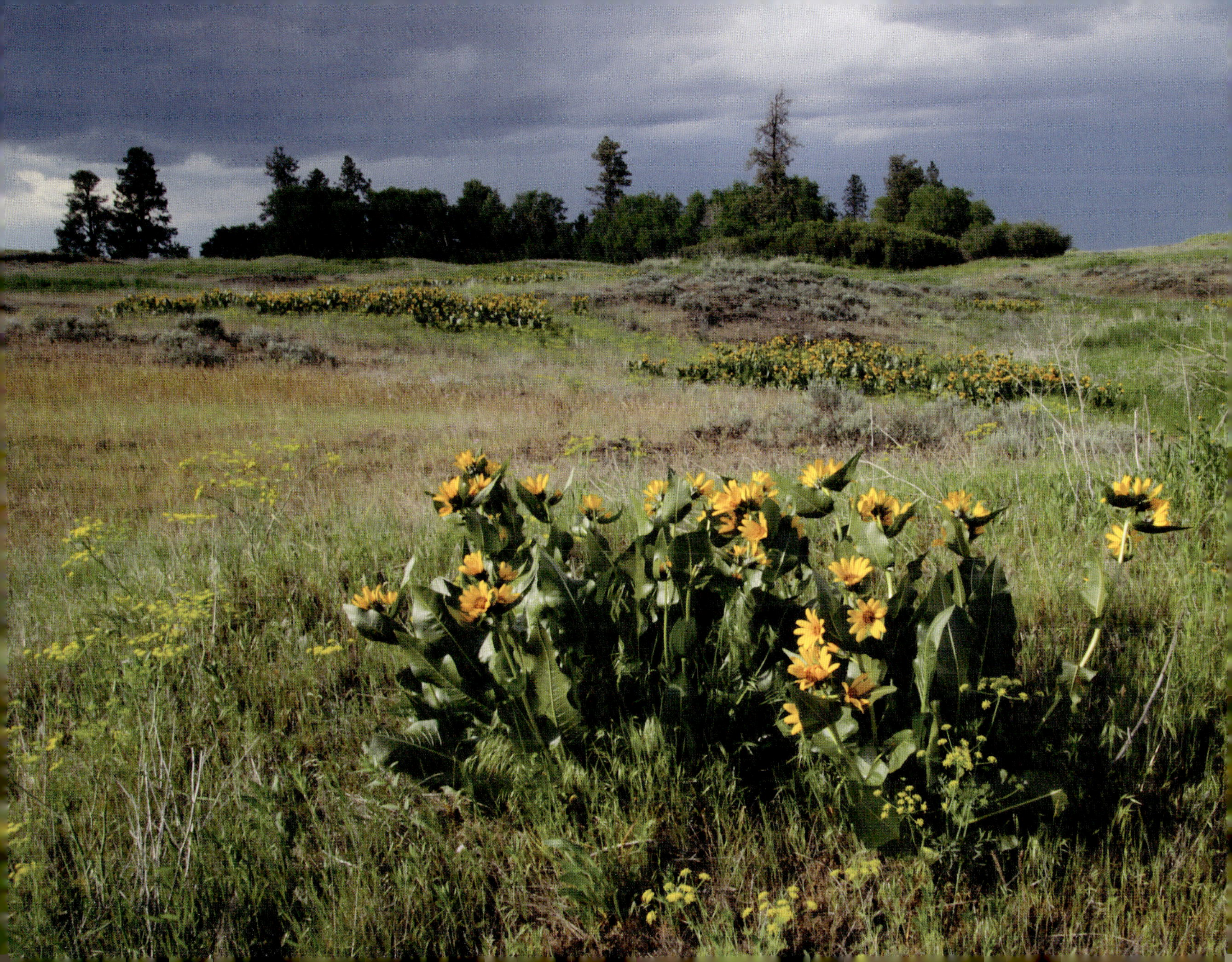

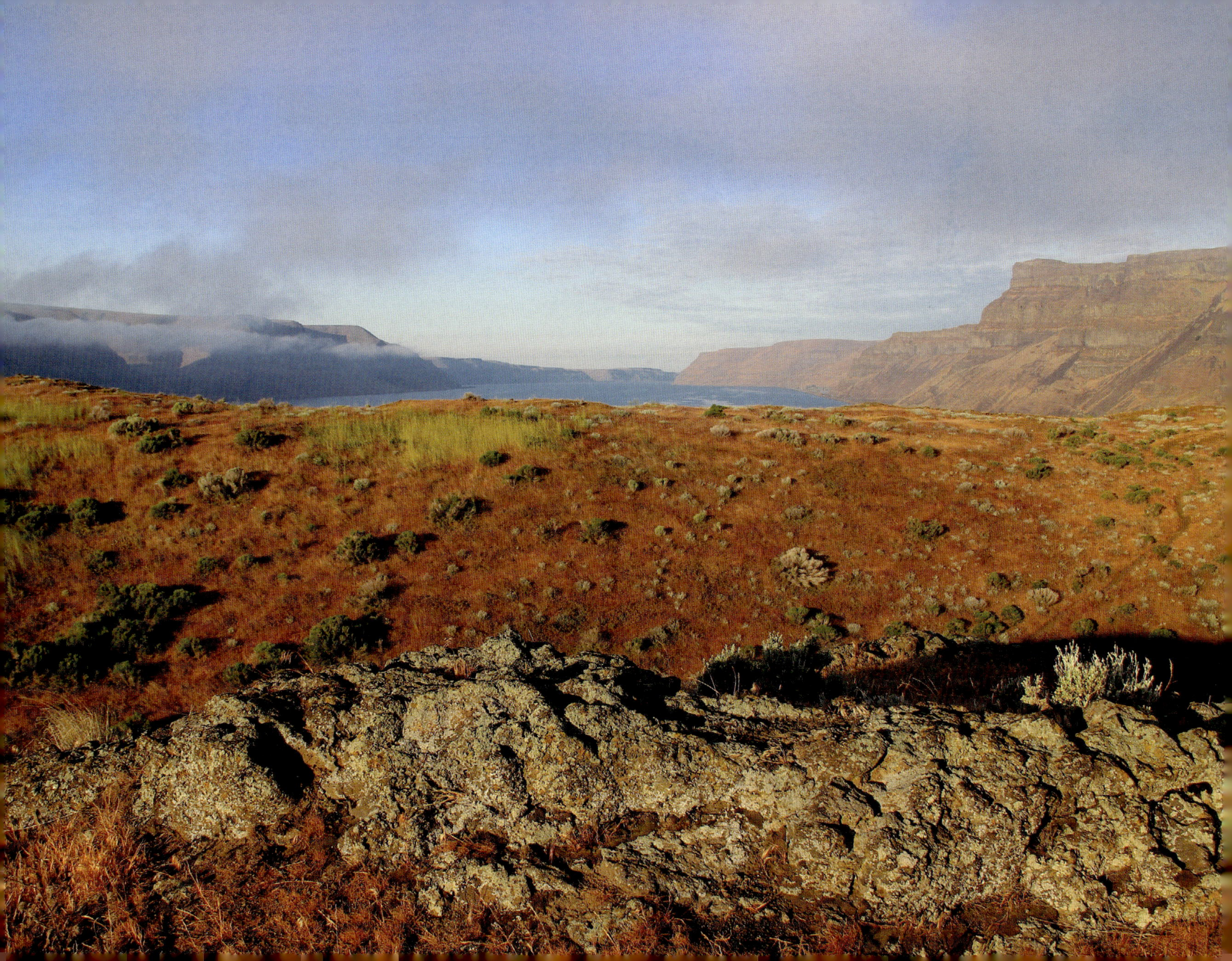

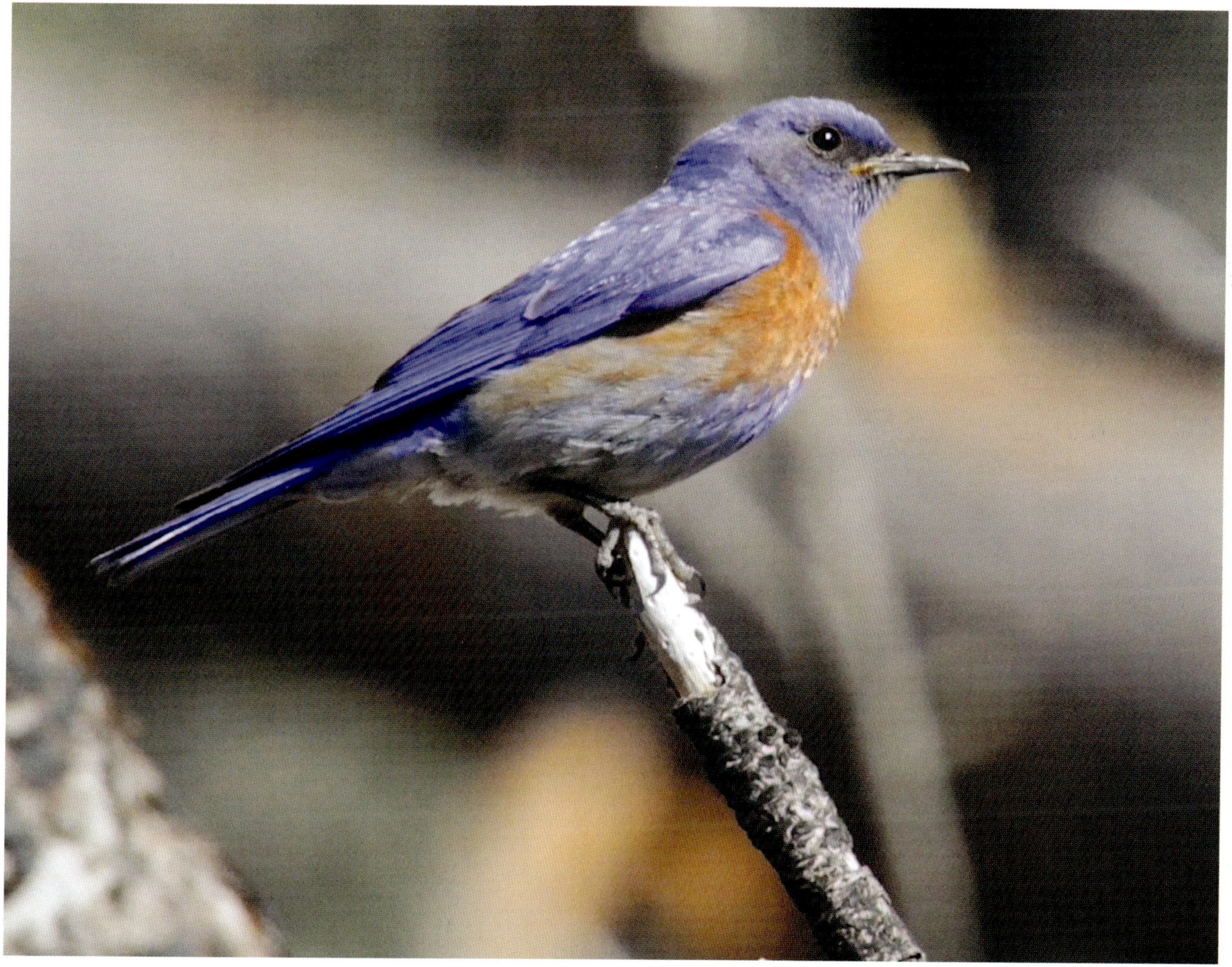

Left: Sunrise at Wallula Gap. *Right:* Western bluebird.

CHAPTER I.

"God's Country," she said.

My grandfather—that's him dancing with my mother, Joan, on the facing page—introduced me to Wallula Gap when I was nine. How purposeful he was about this I don't well recall.

Because I grew up overseas, I would only get to spend time with him for a few weeks every two years. He would usually set aside at least three of those days to take me camping and fishing. For trout and steelhead, he favored the Tucannon River, which flows northward out of the Blue Mountains in the southeast corner of Washington. From where he and my grandmother lived, in Pasco, one of the routes to the upper reaches of the Tucannon is through Wallula Gap, where the Columbia River does a tight U-turn before making its long, straight run to the Pacific. I experience it today the way I experienced it in 1966: awed by how the swells of burlap hills suddenly become towering walls with outcrops of angular, black rock.

A world away, in Panama, my open-air bedroom would take in shrill blasts and bells from locomotives passing only a quarter mile away. In Wallula Gap, though, the horn salvos from locomotives are rounded and burnished by the broad expanse of water. The sound arrives in gentle pulses and has the embracing resonance of French horns.

My grandfather's name was Gilbert "Gil" Hartman. He and my grandmother, Lois, moved to Pasco in 1931, near the height of the Great Depression. He was an avid outdoorsman and civic leader who made his living as a salesman for the Metropolitan Life Insurance Company.

Gil introduced me to nature, or at least to a more sanguine view of nature than I absorbed growing up in Panama. It's not that Panama isn't blessed with the natural richness of its rainforests, coastlines, and tropical islands. Paradise blooms; toucans, trogons, and tanagers flash their wings. It's just that my dad's dad—my grandpa Charlie—arrived in Panama to join a war against nature. It was a siege the American-led Panama Canal Company inherited from the French, whose late nineteenth-century effort to dig a canal had been disastrous. Mosquito-borne diseases killed thousands of workers, and the mountainous, rain-soaked terrain was nearly impassable. The Americans were able to open a canal in 1914, but the mountains are not willing partners in this marvel of interocean commerce. Above and below the waterline, the unsettled earth tries to refill the shipping channel.

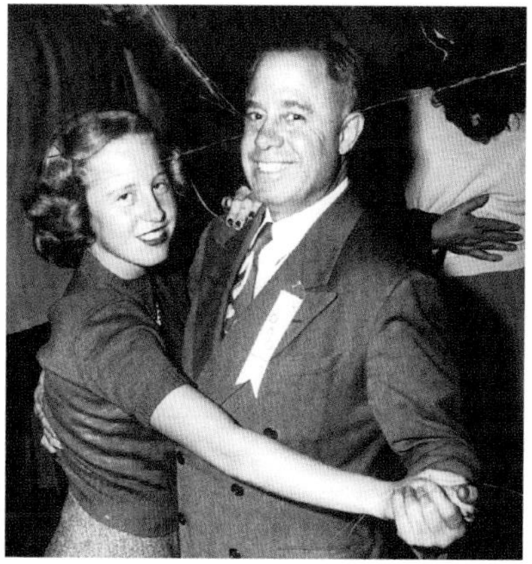

Joan and her father, Gil, circa 1950.

Charlie was a blaster who left New York City for Panama in 1940 and rose to become a drill boat captain in the canal company's vaunted Dredging Division. His job was intense, so he was intense. And his intensity carried over to his parenting. A block from where he docked his drill boat in the township of Gamboa there was a first-class aquatics complex. My dad and his younger brother trained there, between the canal and the rain forest, under Charlie's gaze and booming voice. They became first-rate swimmers. Both brothers received swimming scholarship offers from stateside colleges. My dad chose Washington State and arrived on campus, sight unseen, in 1951. And that's how he met my mother, Joan Betty Hartman, a Washington State cheerleader.

Gil was alert to the hypercompetitive nature of the family his daughter married into. But his interest in sports, especially baseball, was much more about the joy of the game as one of the intrinsic joys of life. His love, like my mother's love, was unconditional, and it was dispensed with élan and grace, laced with whimsical humor and music from their lips.

Wallula Gap requires neither drill boats nor dynamite. It is a beautiful, gaping hole that was chewed through the basalt girders of the Horse Heaven Hills by water—first over millions of years by ancestral rivers, including the Columbia, and most recently by the epic glacial floods unloosed during the late Pleistocene epoch—between 28,000 and 14,000 years ago. The rampaging glacial floodwaters arriving from the north and west not only inundated the gap but completely overflowed high cliffs that are more than 1,200 feet *above* the natural water level of the Columbia.

Today, when I revisit the gap, I feel as I felt when I arrived with Gil a half century ago—as though I'm entering hallowed space in which time stands still, or at least passes in sauntering loops. If there is a desire or need to do so, we can meet the essence of ourselves in these earthen cathedrals. In geologic terms, the foundational basalt at the gap is quite young, a mere 15 million years old—only 200,000 human lifetimes. That says something humbling about human longevity. Yet love borne of human connection is not ephemeral. It can be hereditary and transcend generations. And this is how I experience my memories of this whistling, dancing, fly-fishing spark of a man and the beautiful daughter of his who, as a wife and mother, sacrificed so much of herself. It's hard to find words for this—how the natural grandeur of a place can vibrate with the spiritual inheritance I've received from these fine people. It's the music in the wind.

My mother's kindness is so infectious that it can distract from her remarkable toughness. In that regard and others she is every ounce her father's daughter. On our summer visits to Pasco, my siblings and my father and I would often joke and roll our eyes when she described the Horse Heaven Hills and their sage-covered flanks as "God's country." She was serious. What we saw was a hot, treeless, and sparsely populated expanse, occasionally interrupted by large rivers, such as the Snake and the Columbia, and sporadic stands of poplar, elm, and black locust trees that farmers had planted for shade and wind breaks. But what she saw—in the deepest sense of the word—was home.

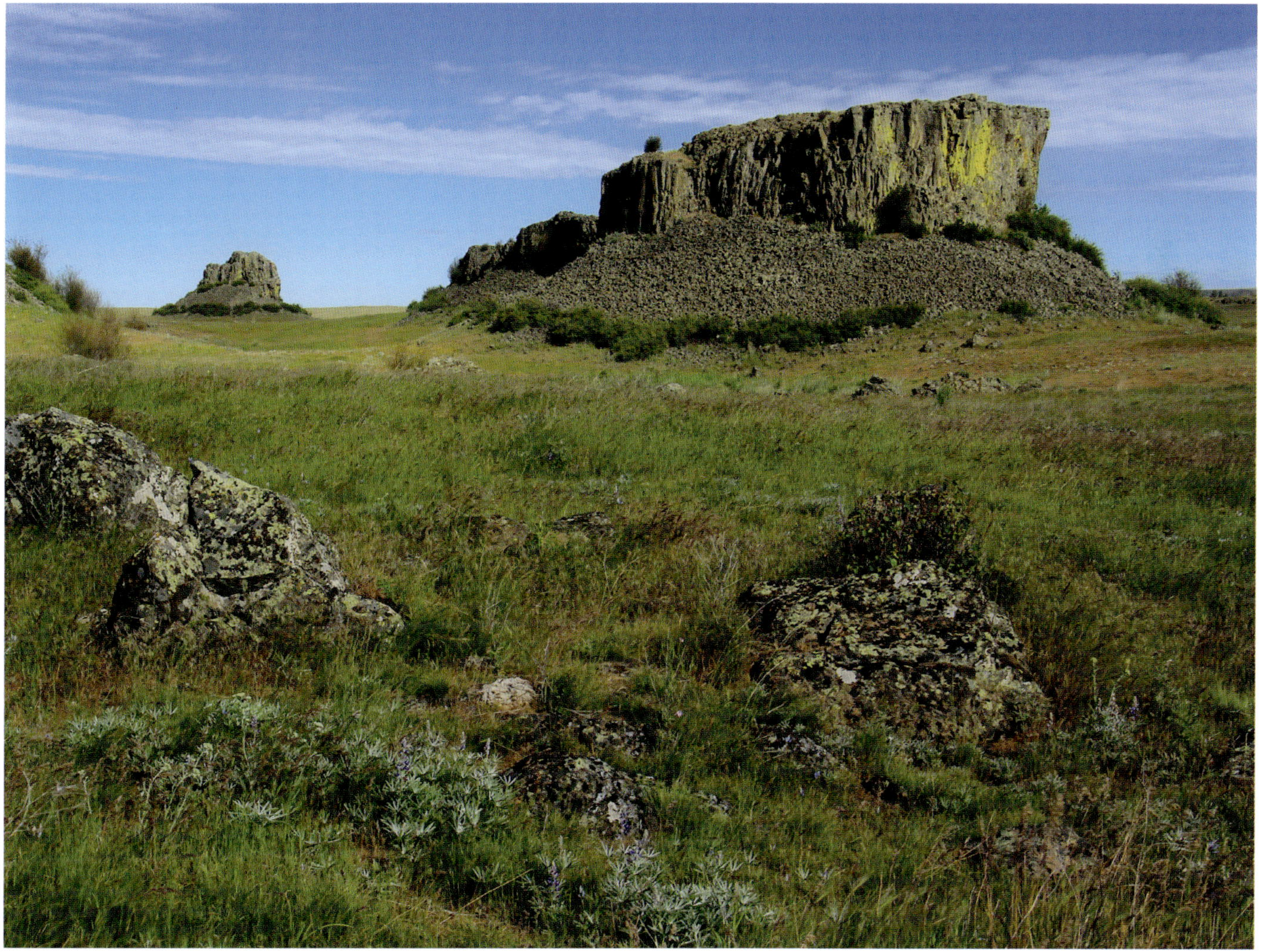

Scabland terrain, western Whitman County.

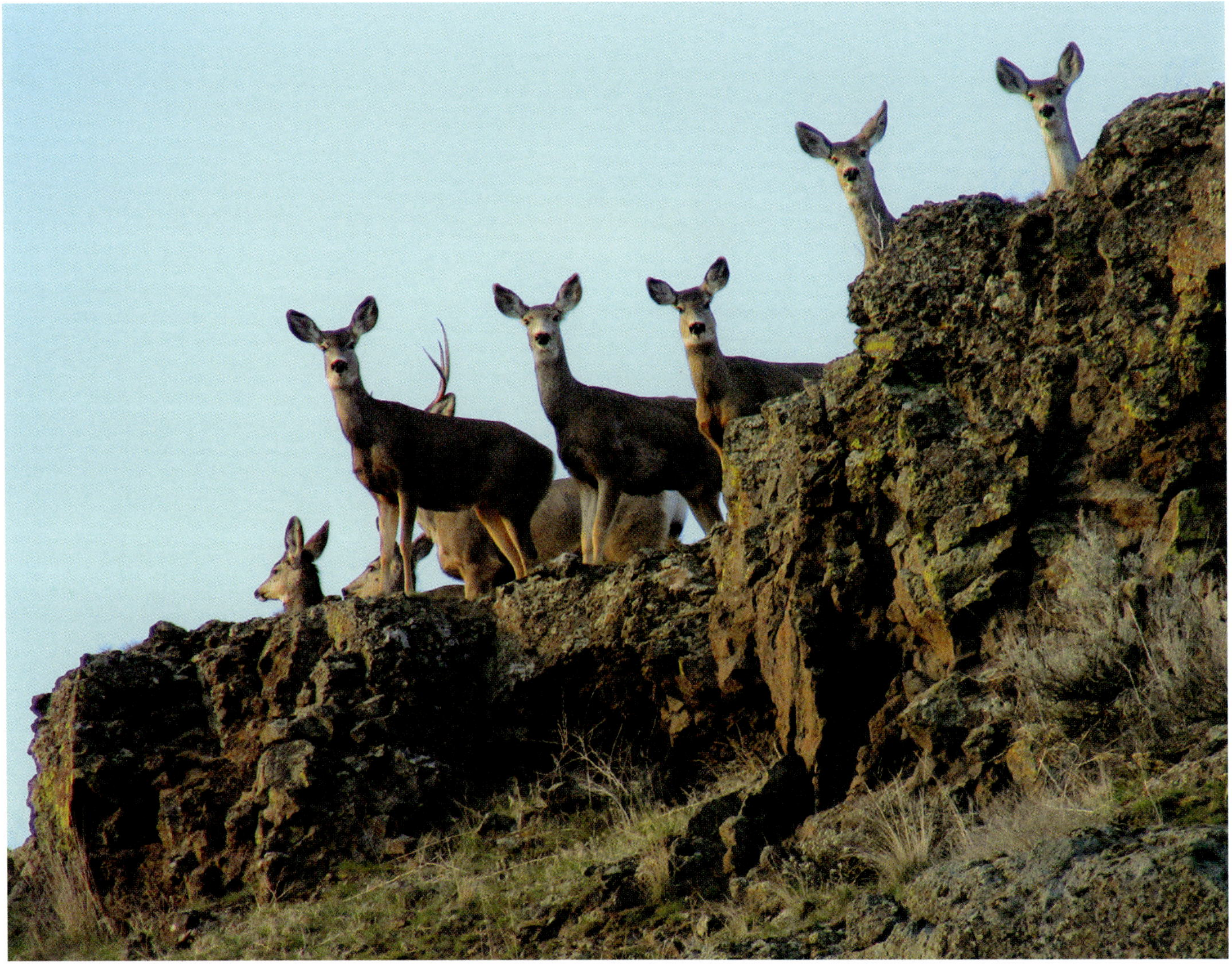

Mule deer in the rimrock, north of Othello.

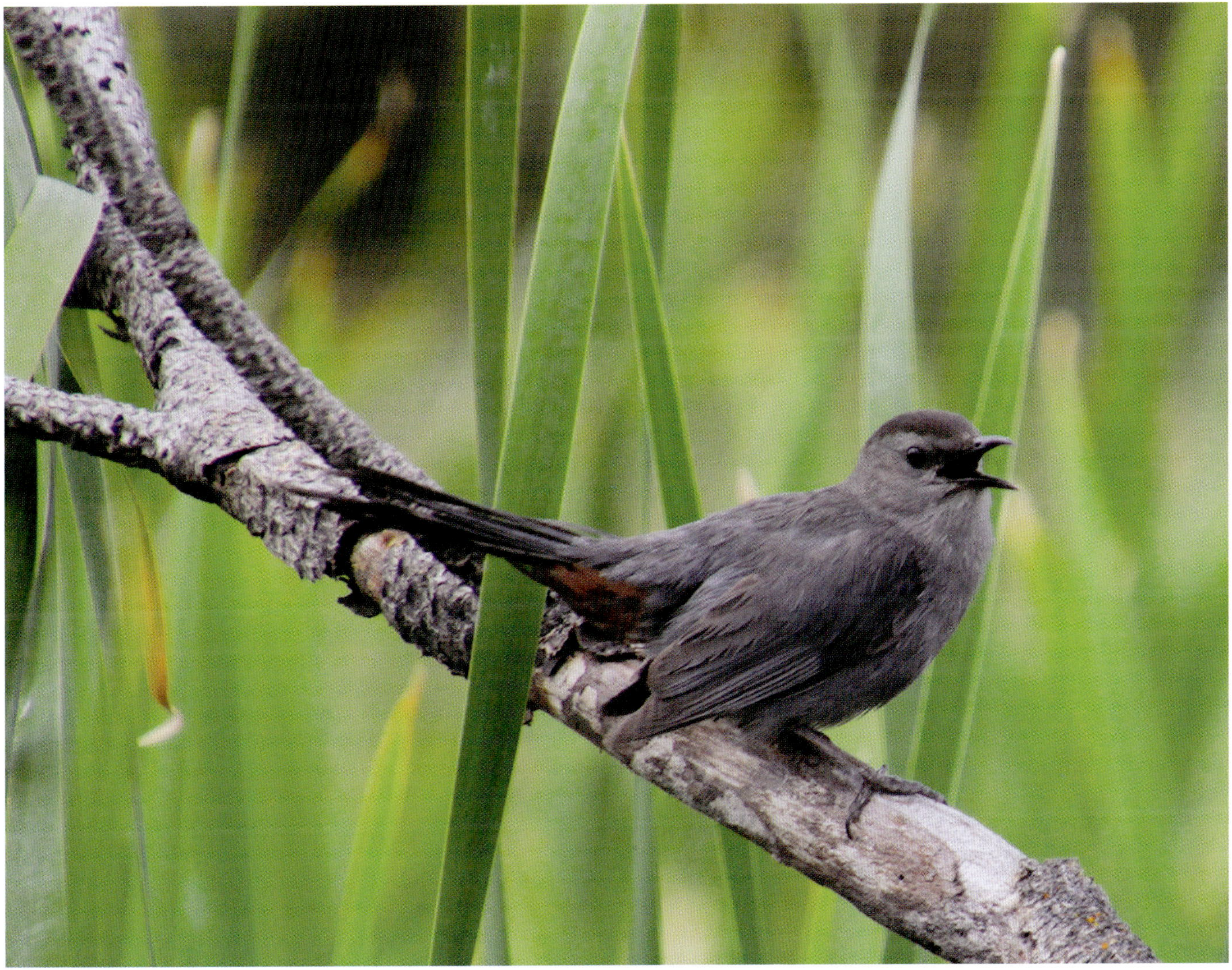

Gray catbird.

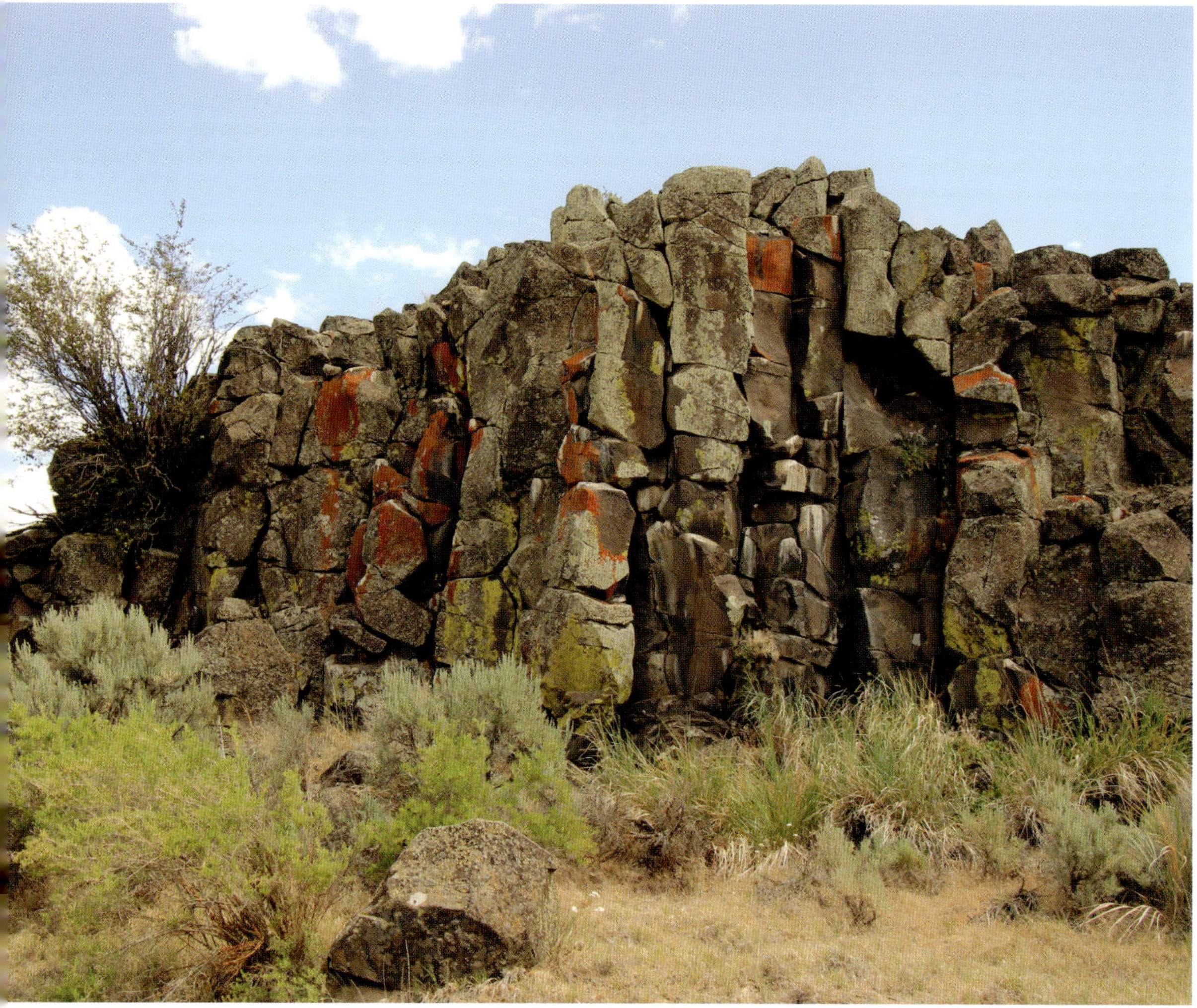

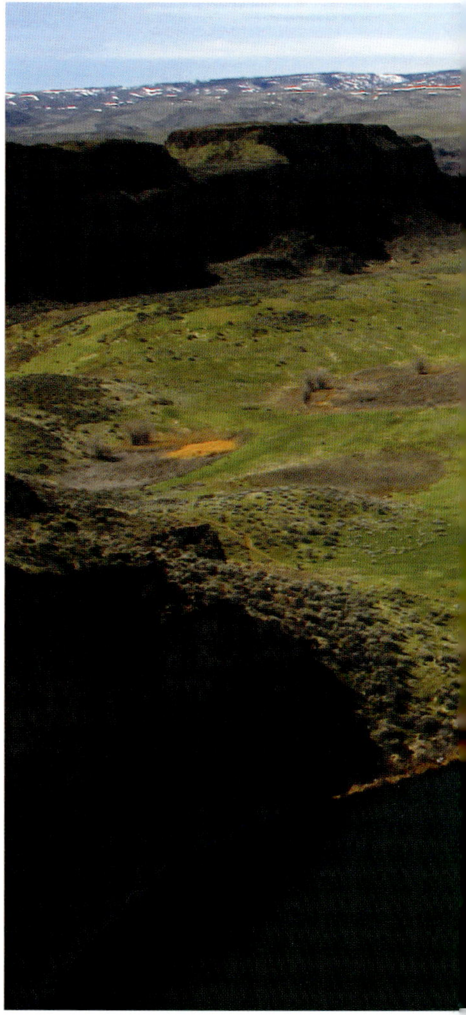

Left: Basalt outcrop, Marlin Hollow, southern Lincoln County. *Right:* North alcove, Potholes Coulee.

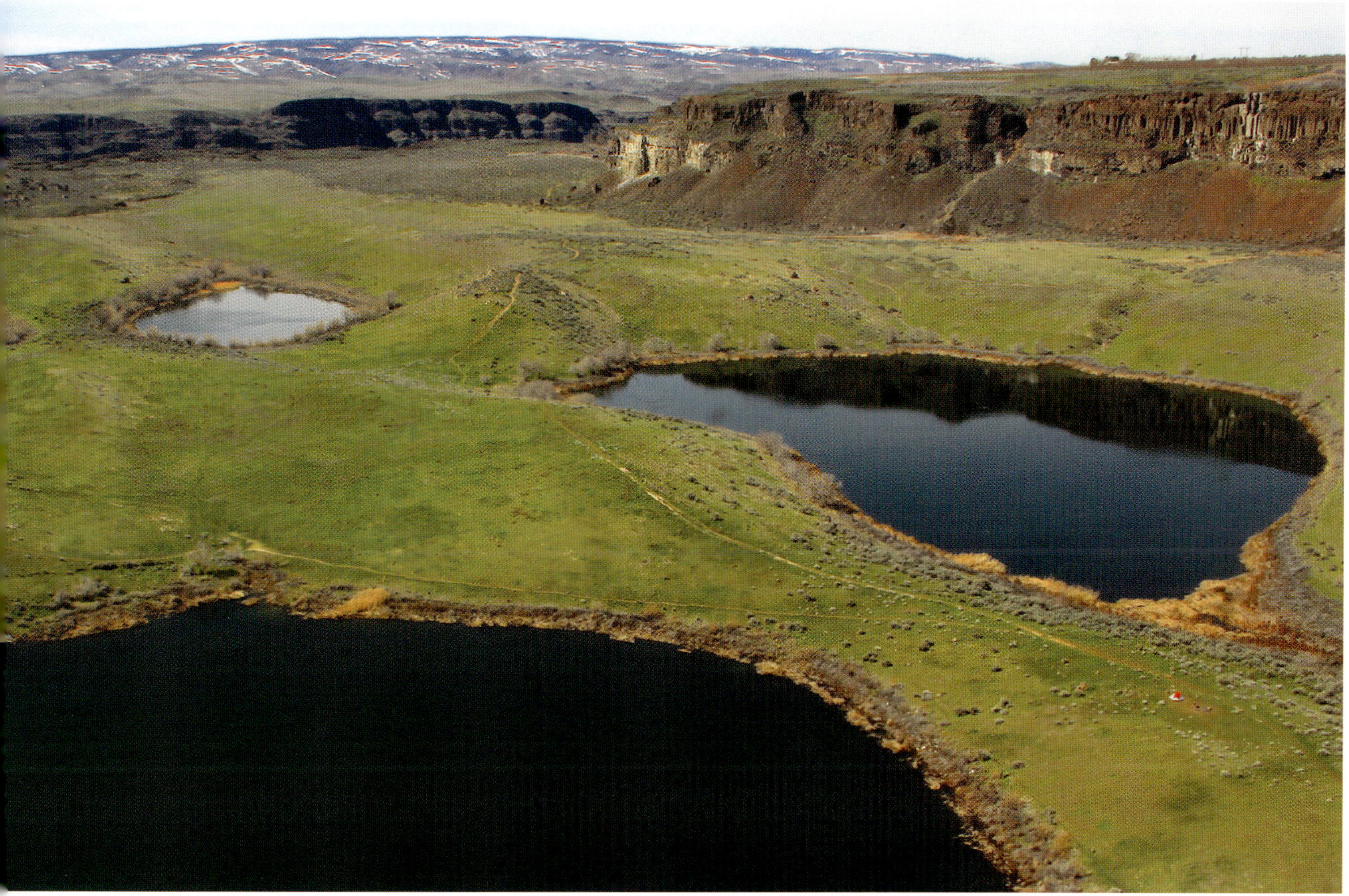

CHAPTER II.

Bretz

Because the terrain in Washington's mid-Columbia Basin rolls gently, if at all, roads in rural Grant County come together like strings on a tennis racquet. Routes named with mere letters of the alphabet align east to west. Roads designated with numbers, some punctuated with decimal points, run north to south. This hyperefficient collaboration of right angles breaks down six miles southwest of Quincy (pop. 6,750), where it looks as though a giant, cloven foot has broken through the carefully gridded earth.

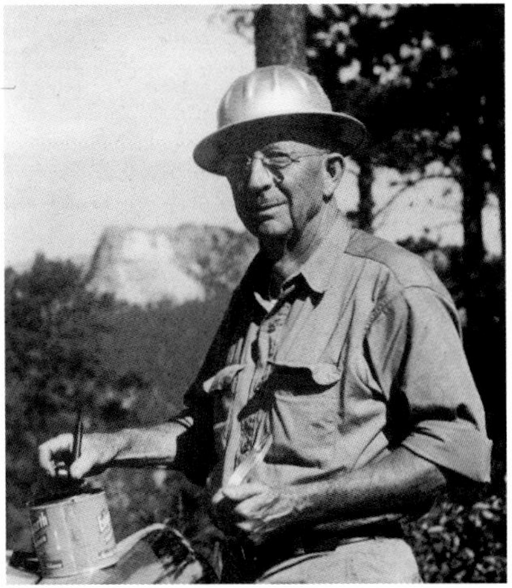

J Harlen Bretz.

This massive, double amphitheater is called Potholes Coulee, or simply The Potholes. A century ago it was both a landmark chasm and a mystery of earth science. And that's why it attracted the attention of a 27-year-old high school science teacher. His name was J Harlen Bretz.

Bretz was living and teaching in Seattle in 1909 when the US Geological Survey published a detailed topographical map encompassing the western edge of what is known as the Quincy Basin. Though it seems unlikely that a map of a well-discovered place would create a stir, this one did. The tightly packed contour lines just east of the Columbia River begged for an explanation as to what event, or series of events, could produce such a gaping feature.

For Bretz, a gifted writer and naturalist, the Potholes enigma stoked an already keen interest in Northwest geology. He formalized his scientific credentials with a PhD in 1913, and within a few years, as a University of Chicago geology professor, began leading field trips to the Northwest. By 1922, his expeditions were devoted to the vigorous exploration and documentation of eastern Washington's scablands.

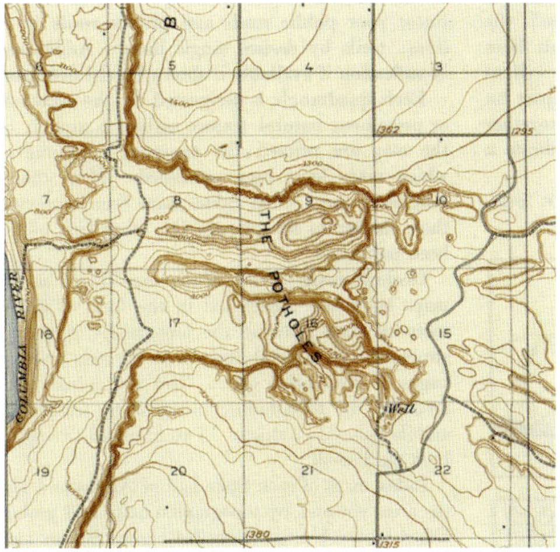

USGS topographical map of Potholes Coulee, 1909.

The Potholes cataract was but one curiosity in a much larger puzzle. No geologist who'd visited the eastern Washington landscape could overlook the evidence of unusual erosional landforms, many of which were spectacular in scale. What set Bretz apart is that he very publicly advanced what was, at the time, an astonishing theory: that the talus-skirted walls at Potholes and countless other coulees and dry channels of the scablands could *only* be explained by sudden, catastrophic flooding on a mind-bending scale.

On account of the field evidence pointing to a colossal water source north of Spokane, Bretz referred to his theory as the "Spokane flood." He did so knowing his interpretation would challenge the prevailing thinking in early twentieth-century geology that even the most cavernous features are created by familiar watershed forces shaping the landscape over long periods of time. And he was surely aware of how his argument would likely chafe scientists weary of the long-standing cultural battle with religious fundamentalists holding fast to literal interpretations of creationism, Noah, and the biblical floods of Genesis.

When I learned about Bretz's remarkable career, two aspects of his story made a lasting impression on me. The first is his (and his students') meticulous summer fieldwork that involved hiking hundreds of miles through the hottest and driest stretches of eastern Washington. The second is the abject resistance he was forced to endure—for decades—from some of the biggest names in American geology.

To be sure, Bretz would not have become a renowned geologist had he not been vindicated. But what made him a legend is the uncompromising quality of his perseverance, the very nature of his hold on the truth, and his commitment to the process of gathering it.

His science is so respected because he was devoted to letting the earth speak for itself. Early on, for example, Bretz took careful note of the bathtub-sized (and larger) specimens of granitic and metamorphic rock that he and others were finding in areas where there are no outcrops of these rock types. His detractors could never plausibly explain how these so-called "erratics" could be stranded, often on hillsides, literally hundreds of miles from their source outcrops and in areas well to the south of where ice age glaciers were known to have advanced.

Bretz surmised that the incredulously displaced "bowlders" were deposited by inland icebergs and were thus artifacts—and additional evidence—of the catastrophic floods that had reshaped the landscape. His hunch would be borne out as the larger story came into view. The granitic and metamorphic boulders had been plucked from their host outcrops by ice sheets moving southward from Canada. These thick lobes of migrating ice would then block rivers—most notably the Clark Fork near the Idaho-Montana border—creating massive lakes. When the lakes burst through their ice dams, the shattered ice became a fleet of ice rafts, carrying the entrapped boulders like so much cargo and depositing them in preposterous places.

In his 2015 book, *Ancient Places*, Jack Nisbet tells the story of the Willamette meteorite, a 15-ton chunk of iron and nickel so large that small children can crawl into its largest cavities. Exhumed a century ago a few miles south of Portland, it is the largest meteorite ever found in the United States.

What the meteorite was conspicuously missing was an impact crater. This is because the Willamette meteorite is the most exotic erratic. Scientists think it crashed to Earth in the Canadian Rockies and became encapsulated in the Purcell Trench glacier, which pushed south to form the ice dam for glacial Lake Missoula. Imagine the spectacle of this extraterrestrial chunk of metal—trapped in a briskly moving iceberg along with more earthly erratics—passing through present-day Spokane on its way to, and through, Wallula Gap, then down the Columbia River Gorge. When it reached what is now Portland, the ice raft bearing the meteorite would have taken a left turn into the flooded Willamette valley before running aground. Tellingly, several smaller, granitic erratics were eventually discovered within a few feet of where the meteorite was found in 1902.

The phenomenon of the ice-rafted erratics—many larger than pickup trucks and some the size of small houses—vividly testifies to the immense power and breadth of the scabland floods. You find a gleaming white boulder on a hillside like the erratic-littered north slope of Rattlesnake Mountain near Richland. There you can easily visualize how deep the water must have been at the time the ice raft melted and deposited the wayward rock. But the erratics were only a part of the evidence that Bretz set before the nation's geologic community beginning in 1919. His core arguments flowed from widely gathered observations about the nature, shape, and breadth of landforms that defied conventional explanations.

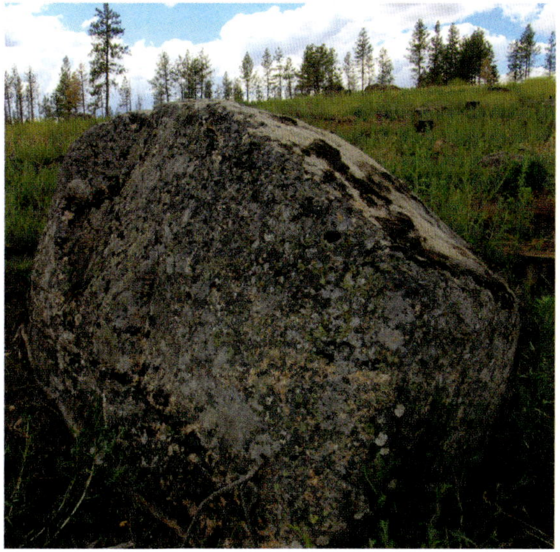

Ice-rafted erratic west of Spokane.

To Bretz's trained eyes, it mattered a great deal that the eroded pathways in the scablands were broad channels, not V-shaped valleys, and that the channels were extensively braided, joining and separating across hundreds of miles. The primary scabland arteries had been filled not just to the brim but had overflowed to create secondary channels, some of which were enormous. The more fieldwork he completed, the more confident he became in his conclusion that only an epic flood could account for his and his students' observations.

Today the Spokane flood is known to geologists as the "Bretz floods." It's a concise homage to Bretz and takes the plural form on account of the evidence that there were at least dozens of the cataclysmic outbursts as the Purcell and Okanogan Lobes of the Cordilleran Ice Sheet advanced southward during the later stages of the most recent ice age. Every writer who takes to this subject confronts the limitations of language because even adjectives like "cataclysmic" are inadequate. The forensics of the geologic research broaden our understanding, yet the science is no substitute for the engaged imagination needed to visualize the power and breadth of what happened. There's nothing else like it in the observed history of the planet. Earth scientists quite literally have to look to Mars to find similar features on the scale of those created in the channeled scablands.

Bretz's legend didn't require an inconceivable theory. Nor did it require a cinematic confrontation, although there is one packed into it. After years of fieldwork and several journal articles advancing his great flood theory, Bretz was invited to the Cosmos Club in the nation's capital in January 1927 to present his evidence at a meeting of the Geological Society of Washington. Notwithstanding the honor of the invitation, what unfolded, in the words of Bretz biographer John Soennichsen, was "a planned attack" subjecting Bretz to "the collective bile of virtually every prominent geologist of his time."

It would take decades to repair what, in retrospect, was a shameful debacle. Bretz wasn't Galileo. He hadn't offended the church by upending the heavens, and he wasn't imprisoned by clerics. Rather, he was rebuked by other scientists for what turned out to be first-rate science. In a sense, he'd threatened their gospels. And while they couldn't lock him up or shut him up, they could impugn his reputation. In the aftermath of the clash at the Cosmos Club, Bretz's life story took on a different trajectory. It would end well but not without a bitter lesson about what a search for truth can inflict upon a searcher. It's a story folded into the scars of a landscape that, metaphorically at least, will always bear the imprints of his well-worn hiking boots.

Right: Clark Fork delta, site of glacial Lake Missoula ice dam.

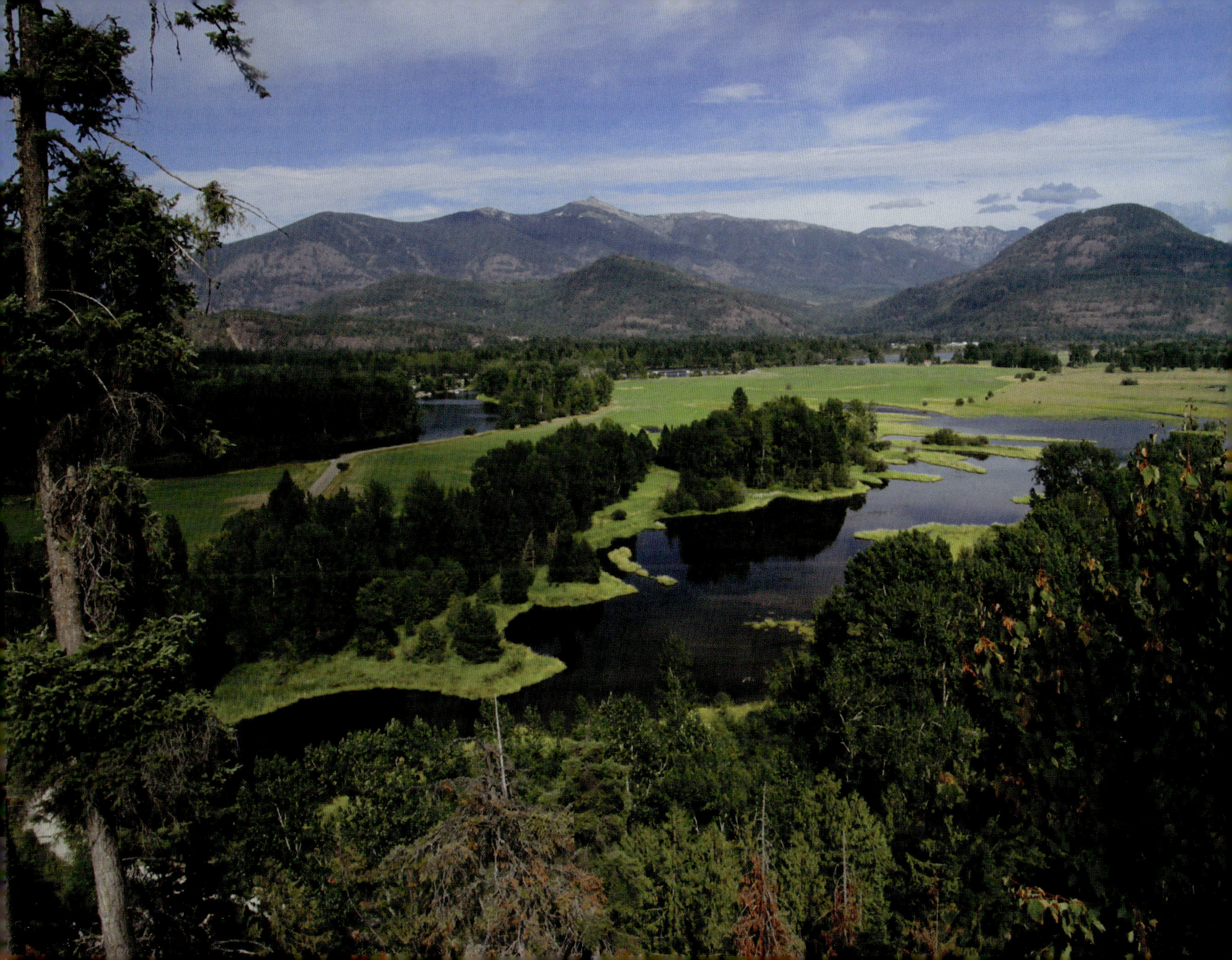

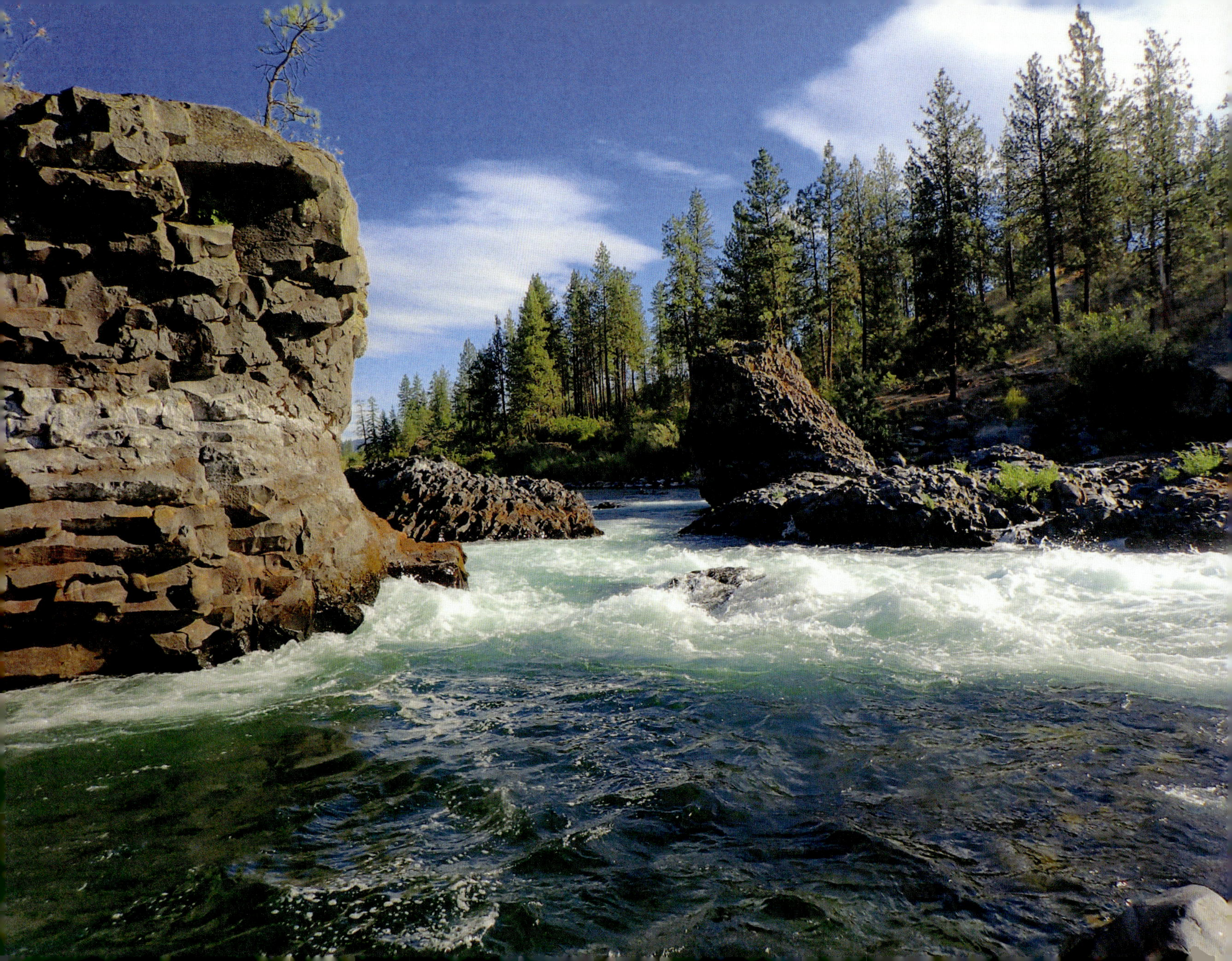

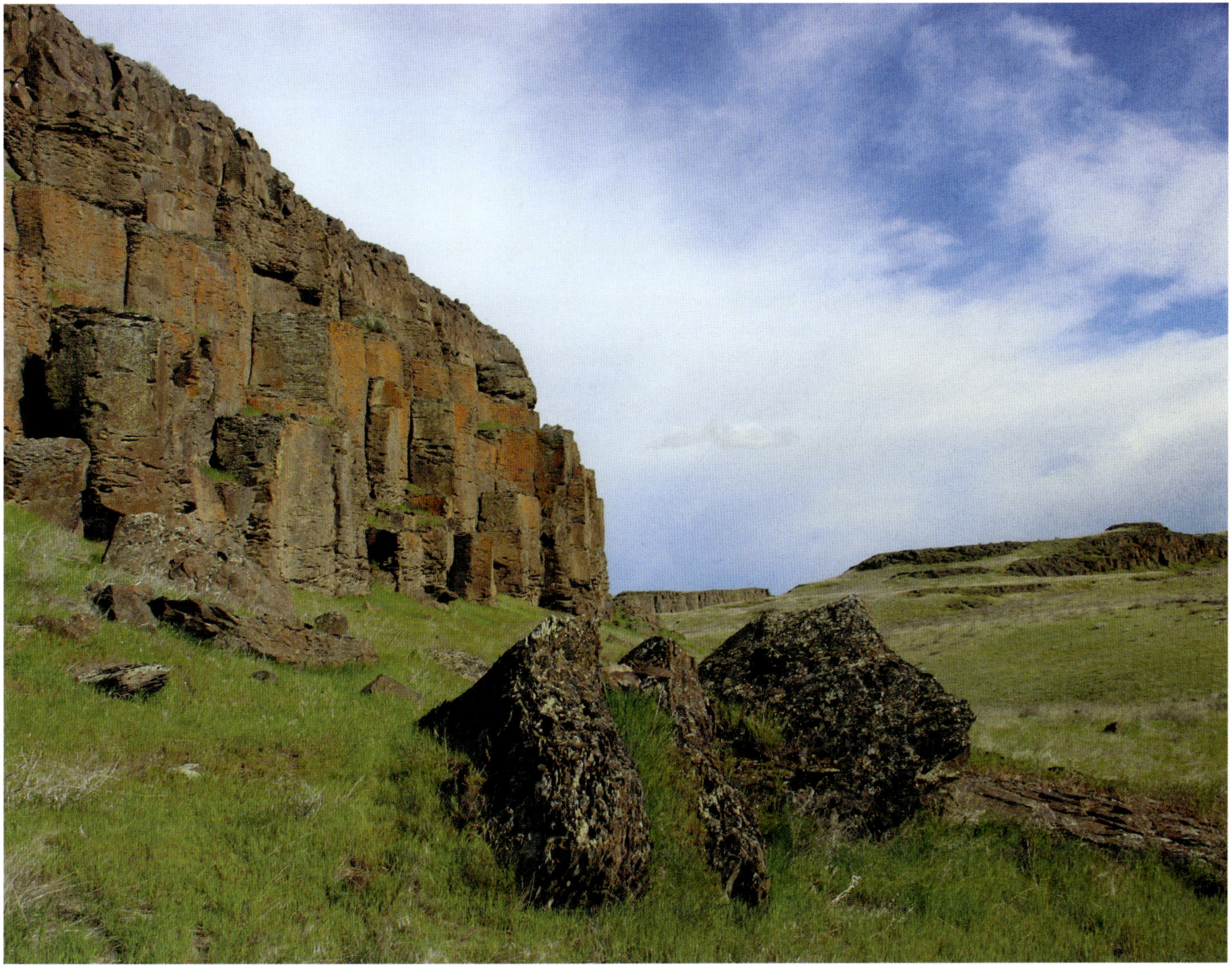

Left: Spokane River at Riverside State Park. *Right:* Ice age flood channel north of Othello.

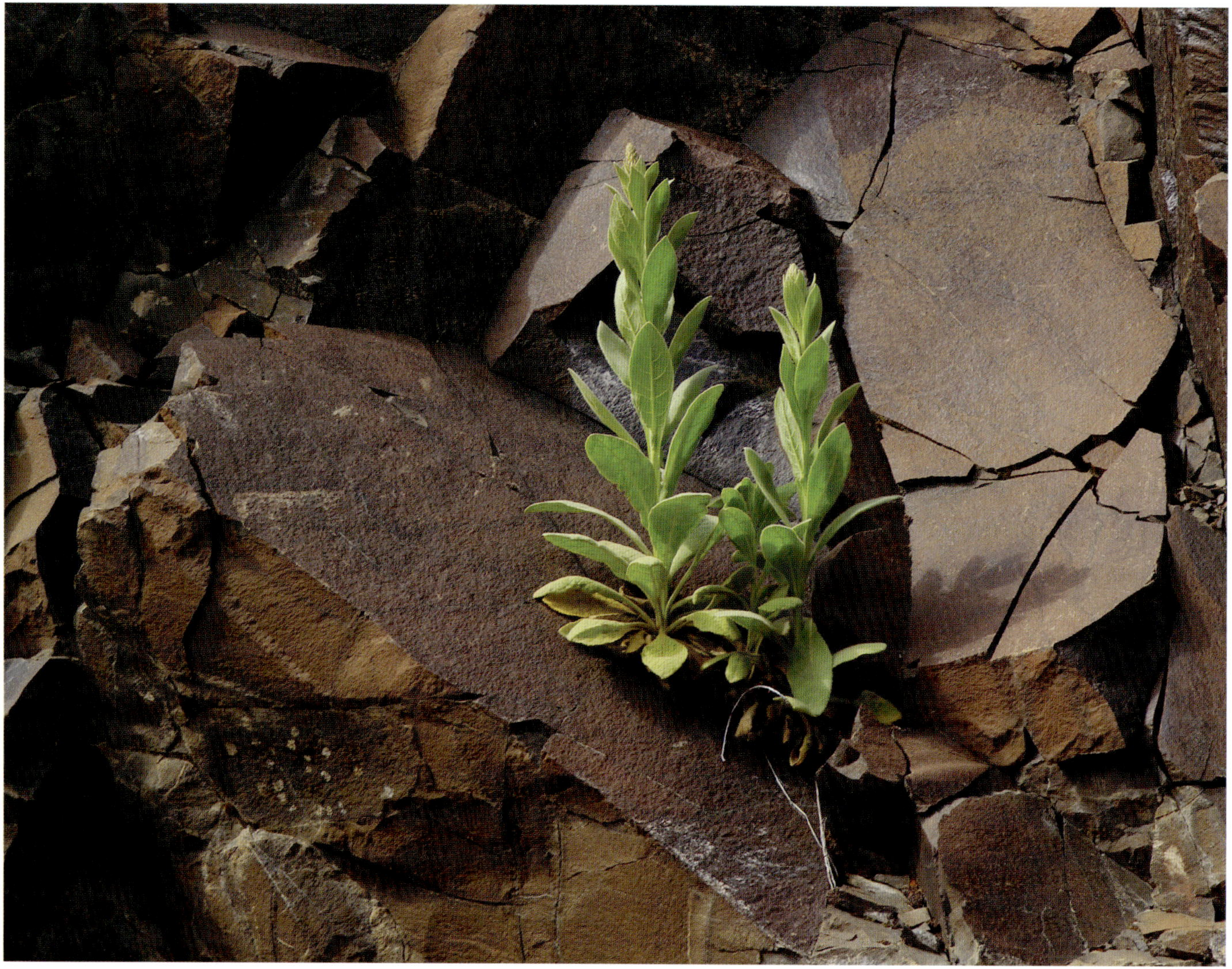

Left: Mullein in fractured basalt, Columbia Plateau trail, western Whitman County. *Right:* Columbia River Gorge at Rowena highlands.

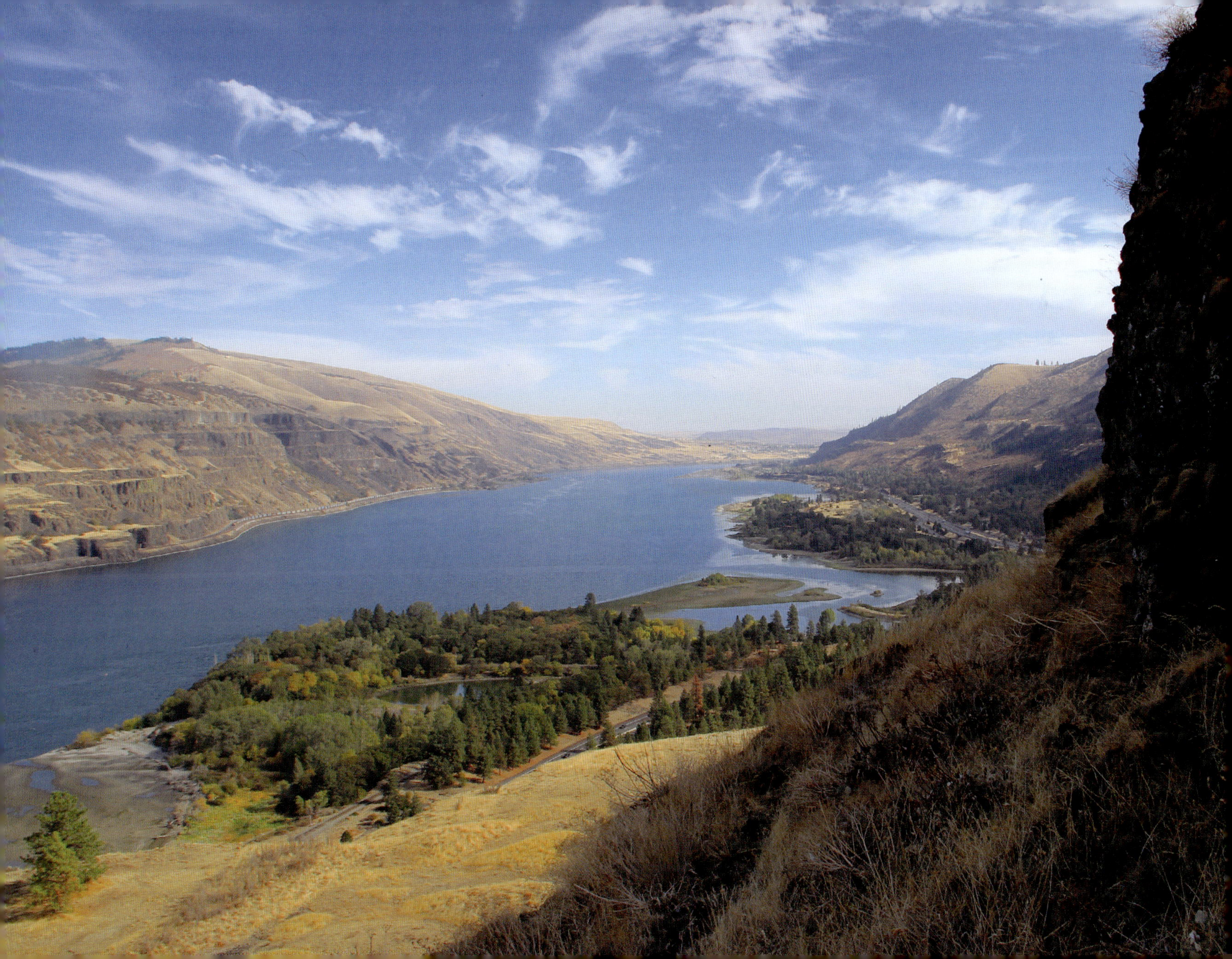

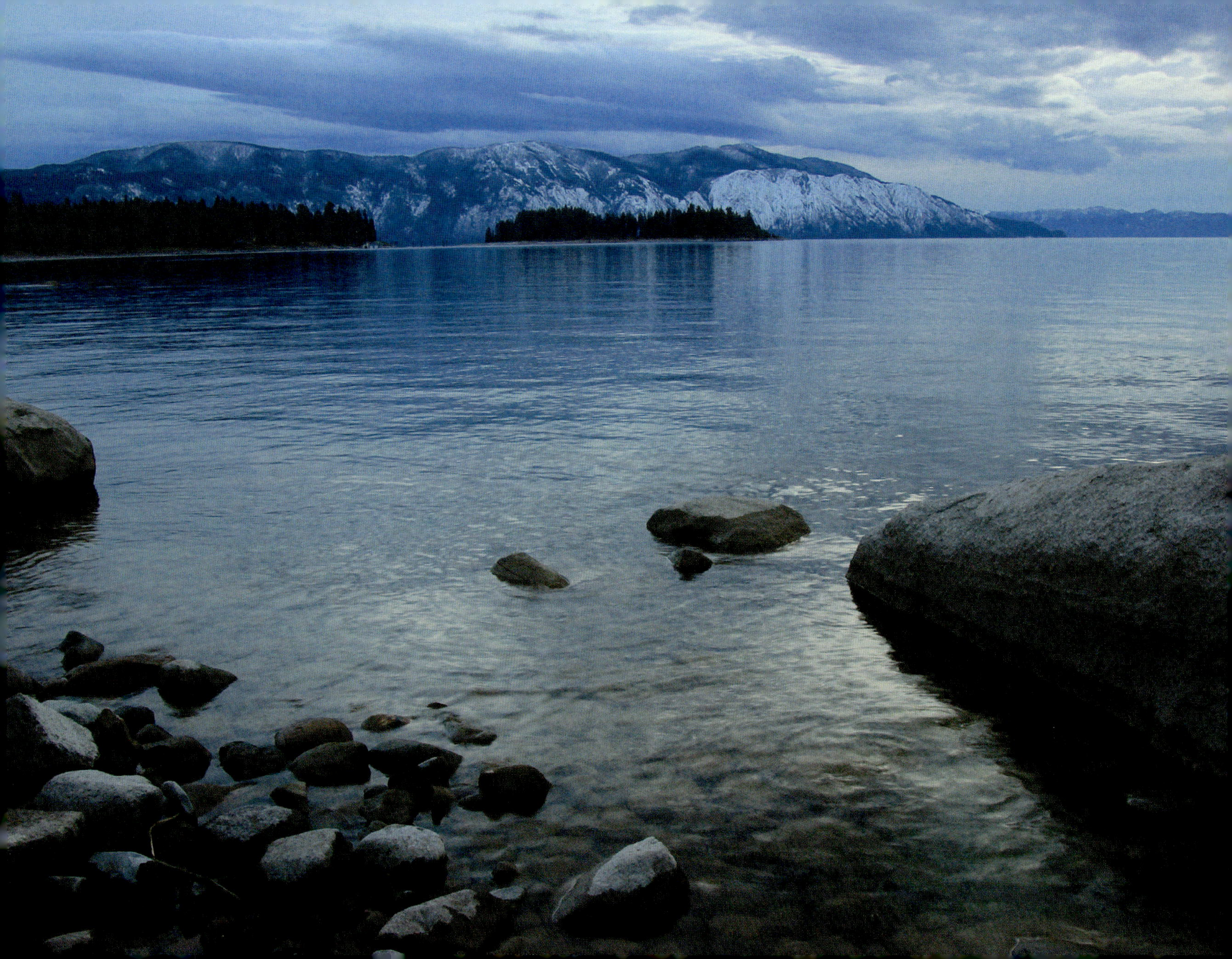

CHAPTER III.

Roaring from the Mountains

If there had been an epic flood—on the near-biblical scale that Bretz had the audacity to propose a century ago—where had the water come from?

Among the earth scientists in Bretz's audience at the 1927 debacle in Washington, DC, was Joseph Pardee. Based in Montana, Pardee had been gathering evidence since at least 1910 that there'd been a sprawling, ice age lake in the northern Rocky Mountains. Laden with glacial meltwater, ancient Lake Missoula covered 3,000 square miles to depths exceeding 2,000 feet.

In some accounts of the gathering, Pardee is said to have whispered to a colleague seated next to him at the Cosmos Club: "I know where Bretz's water came from."

Whether or not the whisper was real, there is correspondence showing Pardee thought the vast depths of the ancient lake might be connected to what Bretz was finding in the scablands. The speculation is Pardee didn't dare raise his voice because his USGS superior, W. C. Alden, was first in line that day to castigate Bretz for his Spokane flood theory.

Thirteen years later, Pardee was among several speakers at a conference in Seattle. There he disclosed something just as mind-blowing as the ice-rafted "bowlders" Bretz had

Left: Green Monarch Ridge from Lake Pend Oreille.

highlighted in his great flood thesis. Pardee had recognized something peculiar about a swarm of low hills on the Camas Prairie northwest of Missoula. He realized the crests on the undulating terrain were actually immense current ripples—evidence that Lake Missoula had drained very swiftly, leaving a stunning watermark on the Montana landscape. When he formally published the finding two years later, in 1942, it served as a tipping point. Pardee's discovery of the giant ripples began to shift the scablands debate in Bretz's favor.

The austere terrain of the scablands seems a world apart from the mist-wrapped, evergreen slopes in the mountains of northern Idaho and western Montana. One of the natural perks of living in Spokane is I can be deep in either world within an hour or so, depending on which direction I choose.

A favorite memory comes from a family outing in 2009, when the four of us traveled north and east, through Sandpoint, Idaho, then continued eastward along the east shore of Lake Pend Oreille. We stopped at Trestle Creek Park on Route 200 near Hope, Idaho, to explore the small beach. My son, Devin, was ten at the time, and before long he grabbed a branch of driftwood shaped like a trident and had us laughing and shaking our heads as he hissed and roared, confronting imaginary foes along the waterline.

Across the lake, we could easily make out the ski runs at Schweitzer Mountain in spring sunshine beneath popcorn clouds. Looking southward, wispy veils of snow were falling. But we could also clearly see what I now recognize as Green Monarch Ridge. There is something peculiar about the way the ridge greets the lake. It is as though someone has used a giant trowel to steepen the slope, so that the face of the ridge drops sharply into the water.

There was another, more obscure clue just a few steps from where Devin was thrusting his makeshift trident into the air. Now, as then, there are large chunks of rocks leaning against one another near the back of the beach.

Colorful and intensely layered, the slabs are among the oldest rocks in the Pacific Northwest. They formed hundreds of millions of years ago as layers of sediment, settling at the bottom of a vast, inland sea. Forged with immense pressure and heat before being pushed to the surface by mountain-building plutons of granite, the rock is extremely hard and

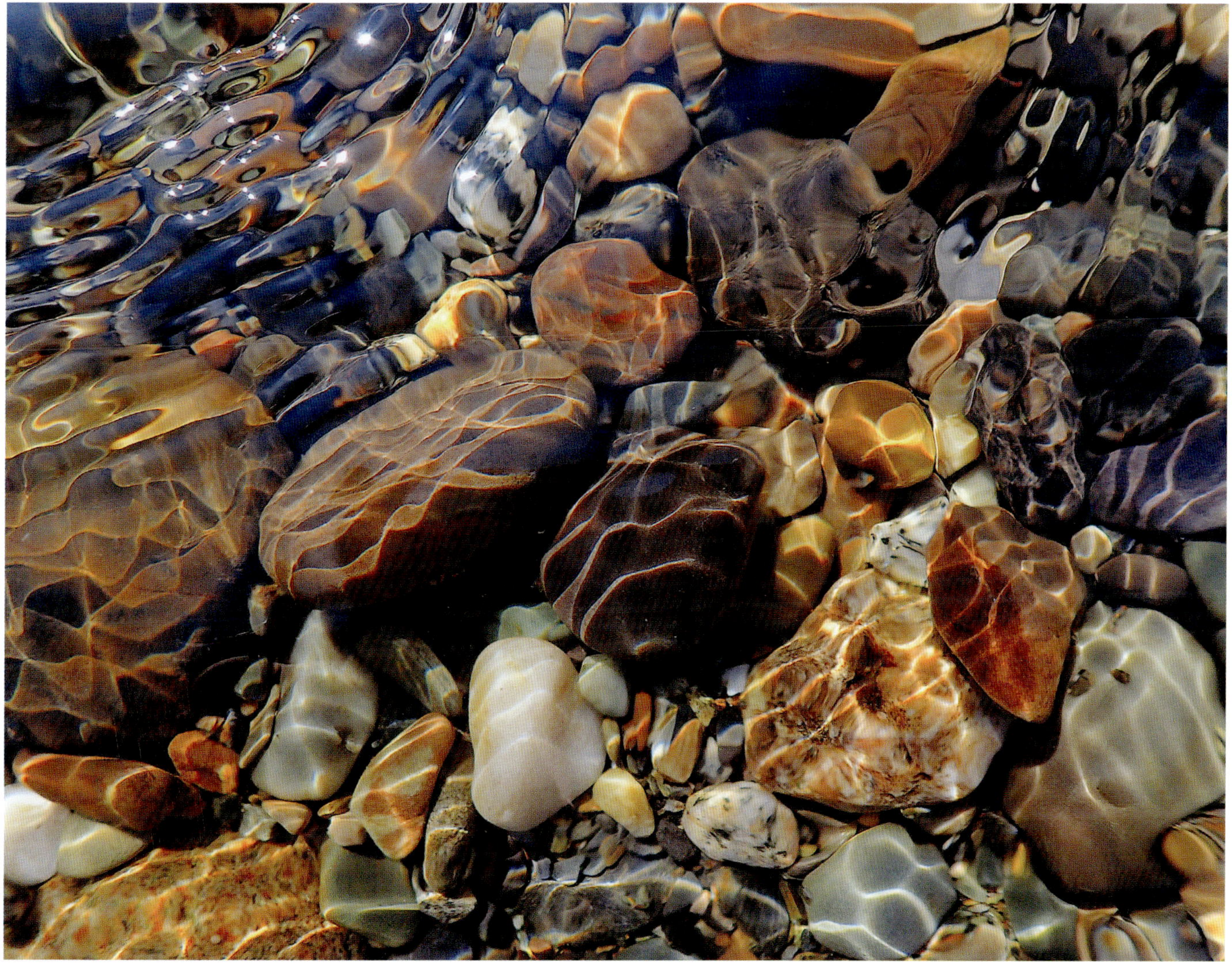

Ice age flood cobbles.

abundant along the Lake Pend Oreille shoreline, including at the face of Green Monarch Ridge, where most of the rock is over a billion years old. There is enough of this very old, very tough, and beautifully layered rock to stop or at least deflect a massive glacier.

And that's a rough, short version of what happened here, just south of Hope, during the last ice age. A massive slug of ice crept southward through the Purcell Trench until it met Green Monarch Ridge. The resulting ice-on-rock collision created a natural ice dam that blocked the Clark Fork River. Without an exit, the river backed up to create the sprawling, ancient lake that J. T. Pardee recognized in his fieldwork. Over time, highly pressurized water would infiltrate and weaken the dam, causing it to rupture with incomprehensible violence and release an amount of water and ice roughly half the volume of Lake Michigan. In a cycle that apparently continued for thousands of years, the lake would quickly empty, but the dam would re-form as the next load of blue ice from the Purcell Trench pushed southward.

What Bretz called the Spokane flood is what happened each time the ice dam exploded. The resulting inland tsunami stormed southwestward across the Rathdrum Prairie at depths up to 700 feet. As it approached the Mica Peak uplift some thirty miles from the breach, the torrent was deflected farther westward, right into and through terrain where the cities of Spokane Valley and Spokane are presently situated.

"Nothing is so firmly impressed on the mind of the visitor to Spokane, as regards its appearance, as the great gorge into which the river falls near the centre of the city."
—Olmsted Brothers report to the Spokane Park Board, 1908

The great floods story in Spokane is somewhat complicated by another glacial lake. For much of the late Pleistocene epoch (ending 10,000 years ago), the Okanogan Lobe of the continental ice sheet would effectively dam the Columbia River a hundred miles to the west. At other times, an ice lobe would block the Spokane River (a tributary of the Columbia) about sixty miles west of Spokane. In either event, the result was the same—the Spokane area would become submerged by a deep lake of standing water, most commonly referred to as glacial Lake Columbia. While there were times when Lake Missoula floodwaters had a clear run through the Spokane area, there were also times when the Lake Missoula floodwaters would catastrophically add to the volume of glacial Lake Columbia. The combined floodwaters would completely overwhelm the Spokane River drainage, including Spokane's signature rimrock gorge, and force breakout floods in several other places—sending earth-shattering waves into the channeled scablands to the south.

The most benevolent consequence of the glacial Lake Missoula floods is largely out of sight. It is the phenomenally deep and long pile of boulders, cobbles, and gravel that the floodwaters left in their wake. This highly porous channel of rock—stretching all the way back to Lake Pend Oreille—is saturated with groundwater that follows the same course as the floodwaters that delivered them. Known as the Spokane Valley–Rathdrum Prairie aquifer, this deep trench packed with billions of cobbles delivers a prolific amount of clean water to a city that receives less than 20 inches a year in precipitation.

In 1969, on the city's north side, a well was drilled to nearly 800 feet. Even at that depth, the borehole in the so-called Hillyard Trough had not reached the bottom of the water-bearing flood deposits. This is a clue to what unfolded at the surface.

Before the Missoula floods, the evidence is that the Spokane River flowed through the north-side trough, to the east of the Five Mile plateau that sits, as an anvil of basalt, on the city's northern margin. With the trough becoming filled with flood debris, the river shifted course to the southwest, to the other side of the plateau. This diversion is what brought the modern river to the city's signature waterfalls and picturesque gorge that the famous landscape architectural firm, Olmsted Brothers, recognized and praised in its landmark report for the Spokane Park Board in 1908.

The gorge itself is a massive flood feature that essentially masquerades as a river canyon. One way to appreciate its incredulous scale is to visit the Centennial Trail trailhead above the popular Bowl and Pitcher rock formation that rises from the riverbed in West Spokane. Although you are 200 feet above the river at the trailhead, you are only midway down the terraced basalt rimrock. Behind you, as you look down at the river, is a massive wall of basalt that rises another 200 feet.

To envision what it would have looked like on a day when Lake Missoula floodwaters arrived, you have to imagine water overflowing not just the cliff below you but the cliff behind you. From there, the escaping water headed southwest, where—six miles from where you're standing—it left giant current ripples on the West Plains near what is now Fairchild Air Force Base. From there, the torrent went to work digging out the channeled scablands as it made its way to Wallula Gap and beyond.

My most intimate relationship with the Spokane part of the Spokane flood is as a swimmer. My swimming hole is in a broad reach of the river, upstream of the Bowl and Pitcher, that cuts through an enormous flood bar of cobbles and boulders left by the ice age floods. The reach is populated by vast schools of northern whitefish and lesser numbers of crawfish and native redband trout. For some reason, the brilliantly colored trout are sometimes approachable and will tolerate me swimming beside them as they hunt. I splash into this realm on a near-daily basis from late spring to early fall. As I swim the channel, I look down upon enormous boulders of granitic and metamorphic rock. Especially in sunlight, they appear as natural monuments and remind me of the cataclysm that brought them here. On what would have been my father's 86th birthday, I swam upstream to the rapids where the trout find their meals in the current. And that's where I spread his ashes.

Right: Ice age flood–deposited boulder, Spokane River.

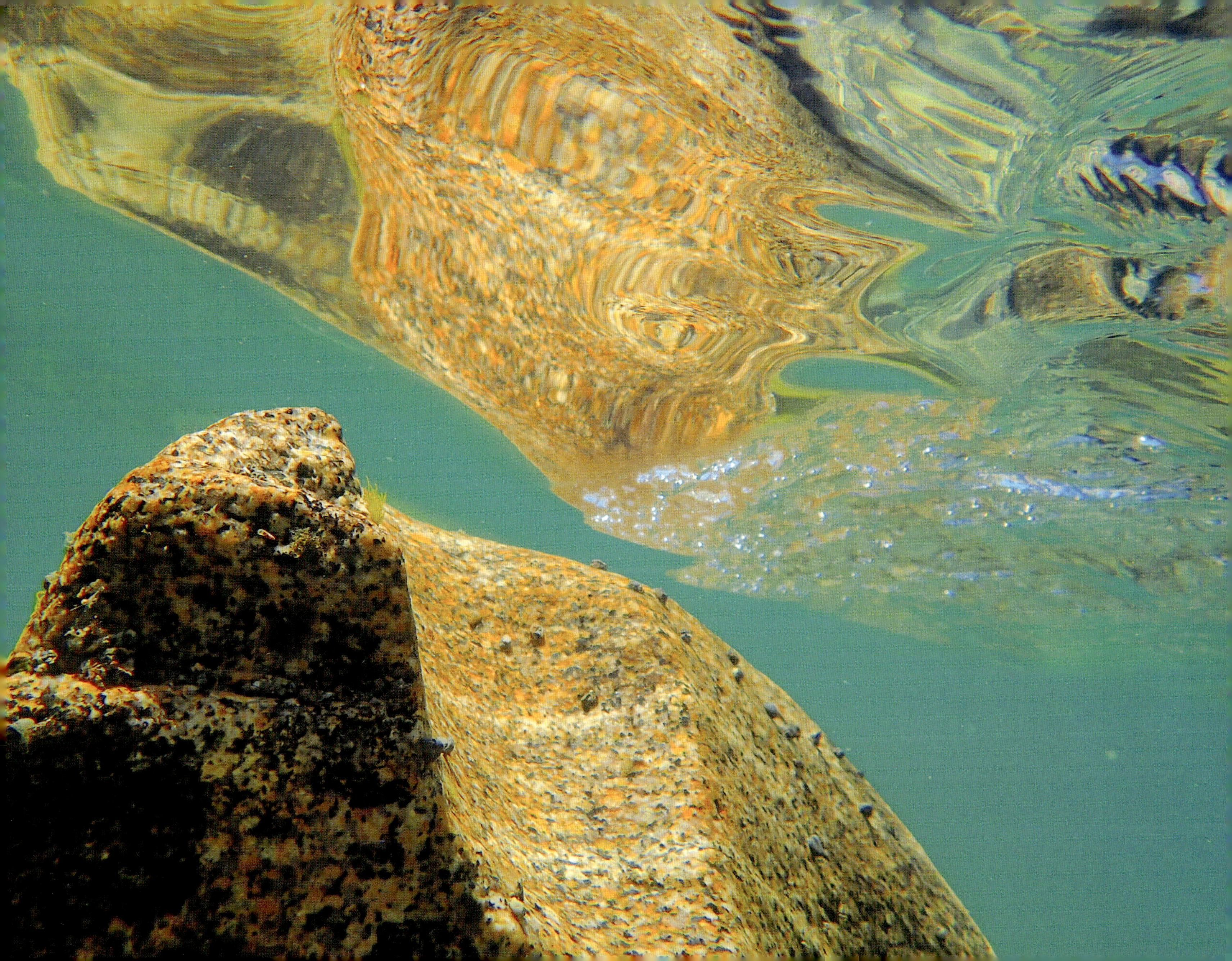

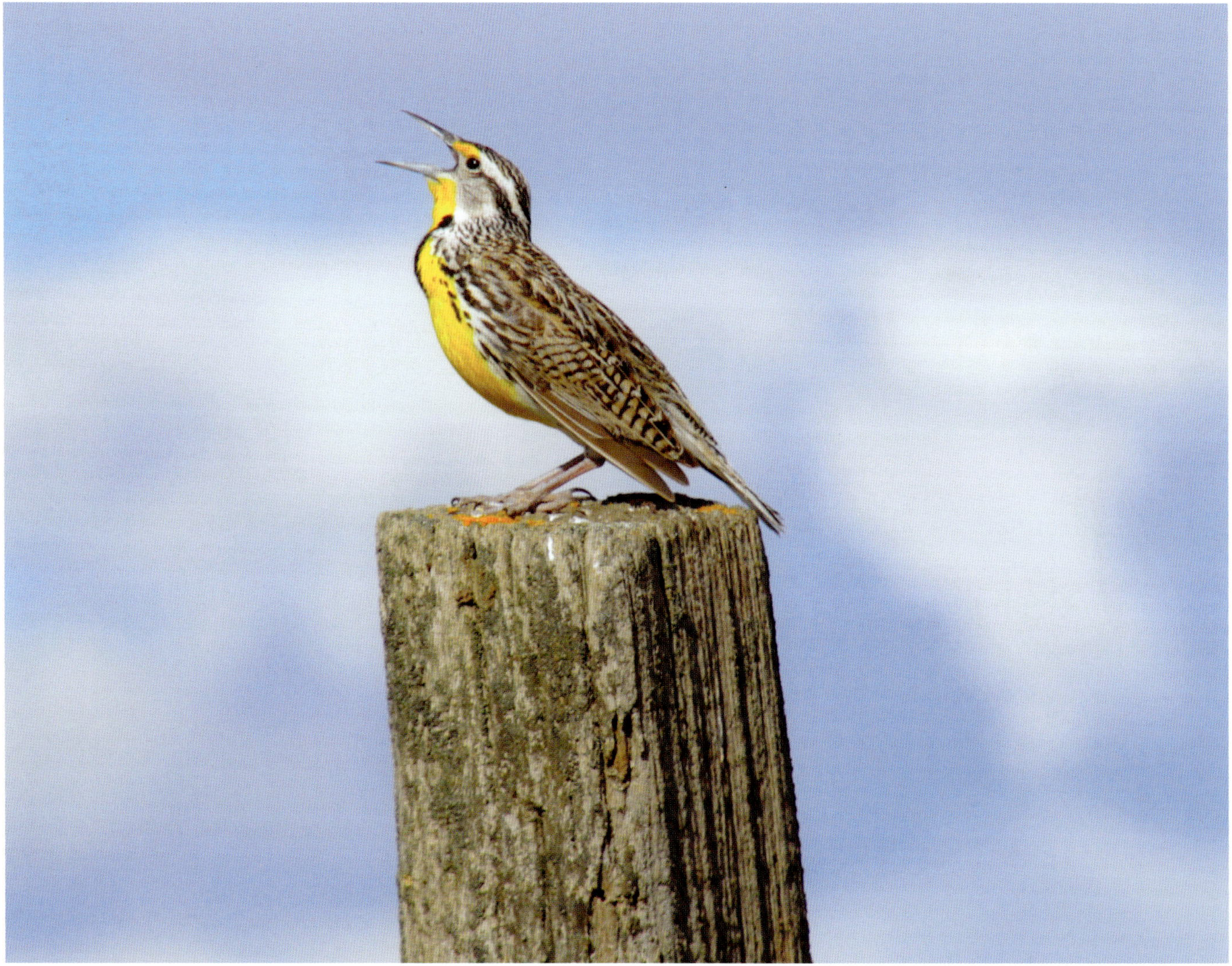

Left: Western meadowlark. *Right:* Crab Creek, Adams County.

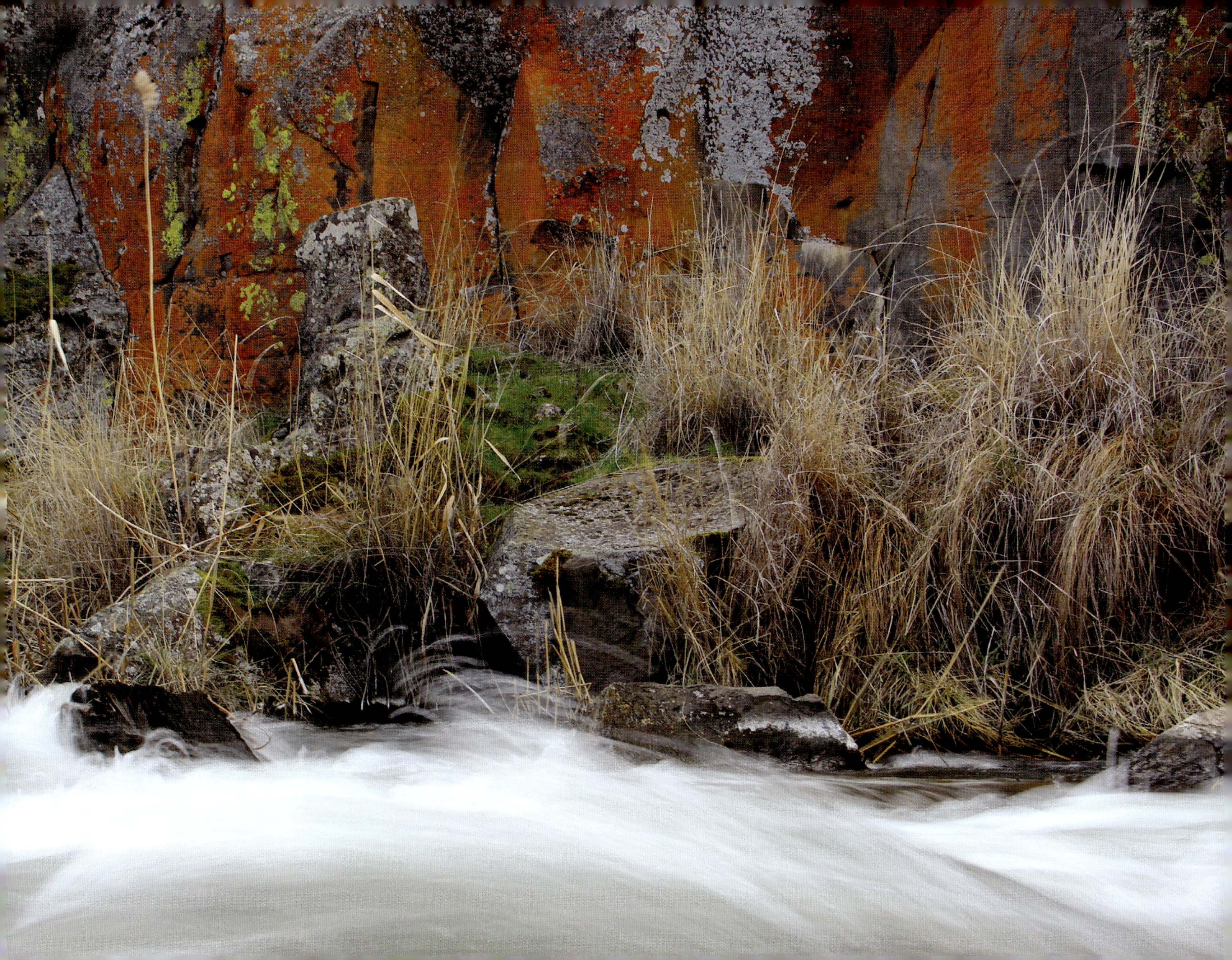

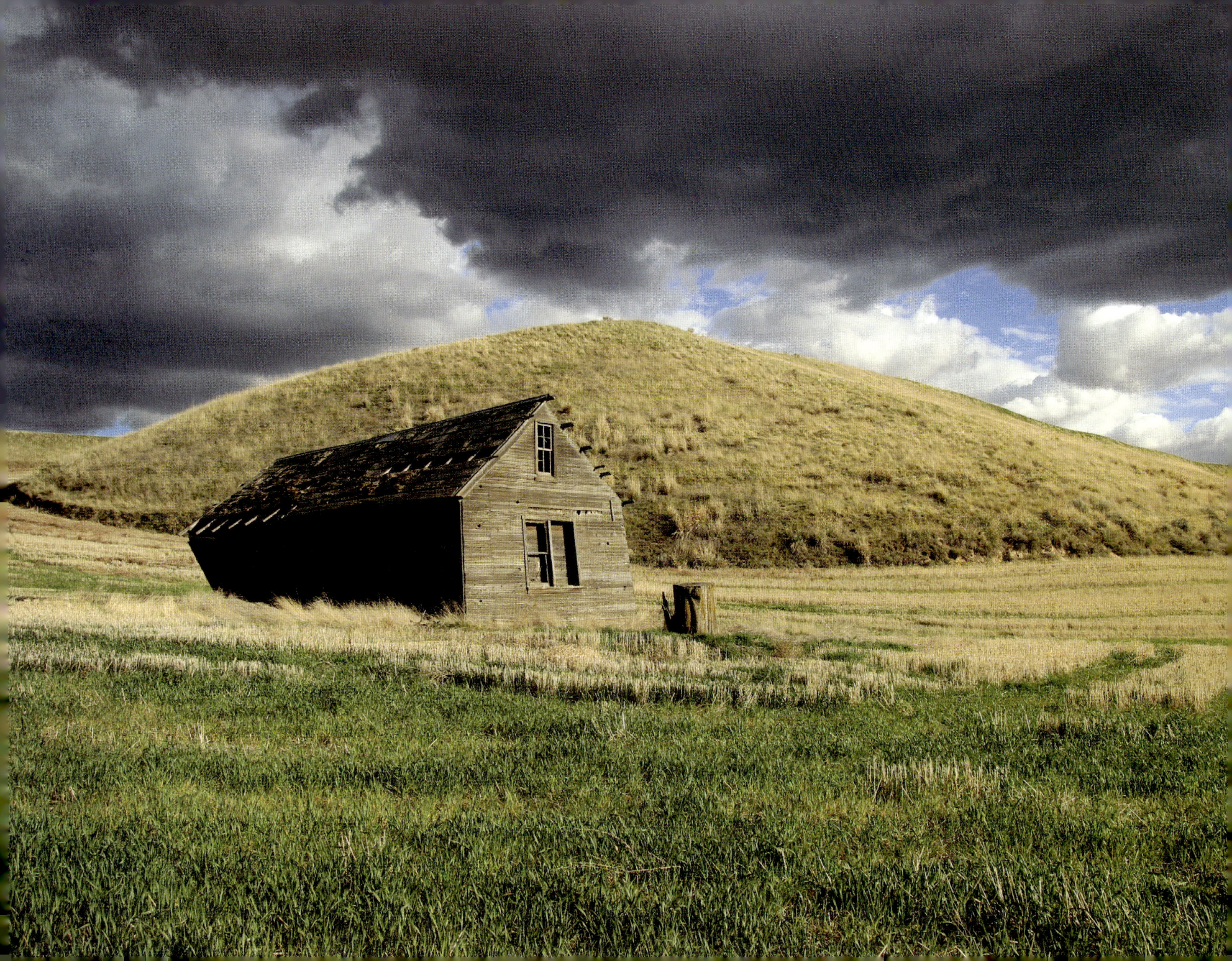

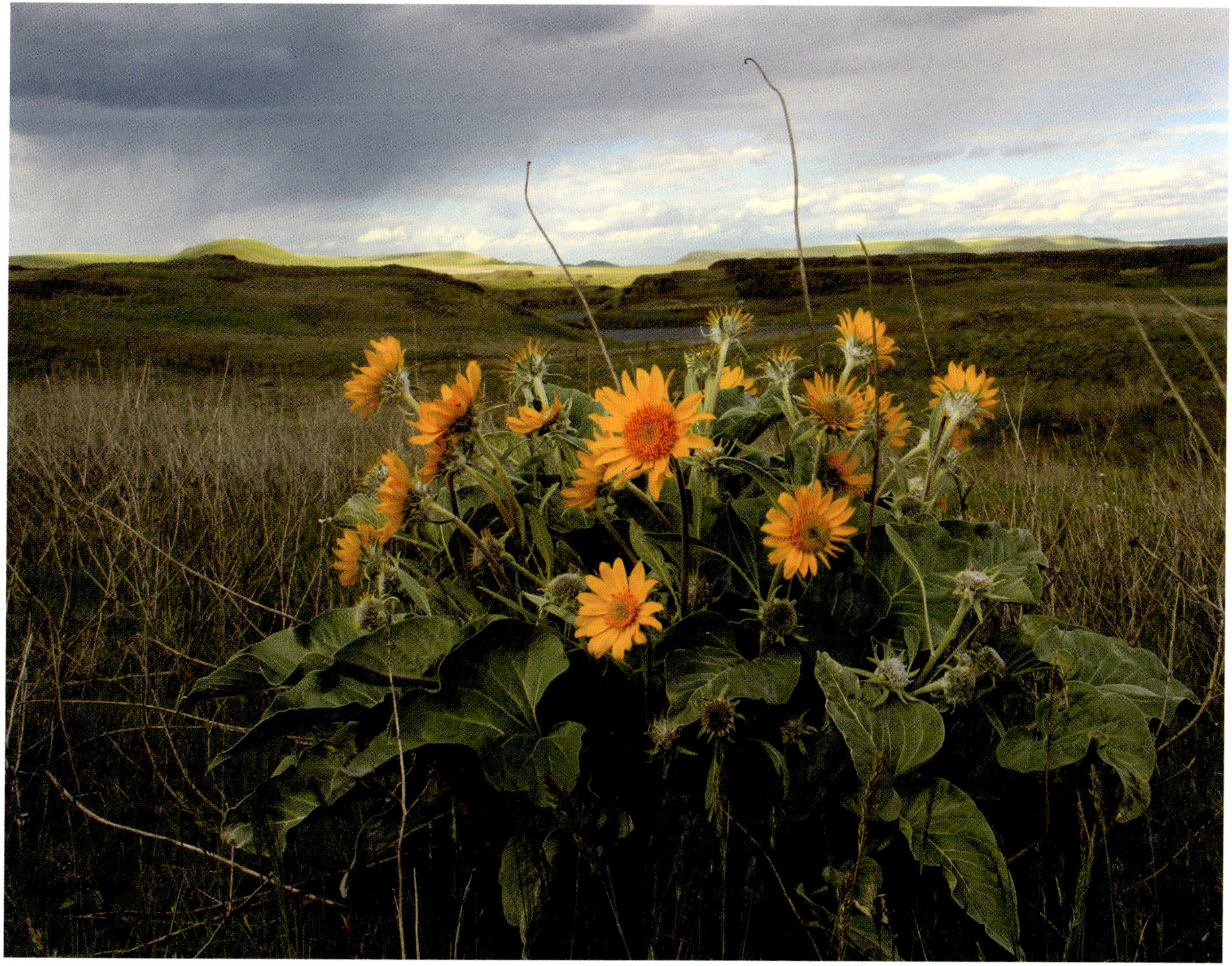

Left: Bretz hill and abandoned farmhouse south of Lamont. *Right:* Arrowleaf balsamroot, western Whitman County.

CHAPTER IV.

The Light behind the Storm

Not all the Lake Missoula floodwaters that overwhelmed Spokane in the late Pleistocene made their exit by surging over the southern rimrock. But the liquified riot that made off in that direction—leaving a rippled watermark on the West Plains—was headed at highway speeds toward Pasco. In less than an hour the stampede of ice, water, and rocks would have overrun the location now hosting the bucolic farm town of Lamont (pop. 77) on land that remains something of a living monument to the story Bretz was trying to tell a century ago.

Gathering its name from a railroad executive, Lamont was incorporated in 1910. As you arrive from the west on Washington Route 23, you can see why Whitman County is famous for its dryland wheat farms. The town is framed on its eastern flank by the fertile dunes of flour-like loess that define the Palouse region of eastern Washington and much of the Idaho panhandle.

What's peculiar is how abruptly the Palouse hills end just short of the main street, Lamont Road. As it leaves town in a southwesterly direction, the asphalt road reduces to gravel and changes its name. Still, it continues for ten miles as a precise boundary between a broad shoal of Palouse hills and the heavily scoured terrain to its west. With a mere turn of the head, your view changes from one dramatic landscape to another—from the graceful lines of the Palouse to the exposed rock, potholes, and general chaos of the scabland.

In his 1923 paper for the *Journal of Geology*, Bretz presented a detailed map illustrating the "Channeled Scablands of the Columbia Plateau." It shows, among other things, how Lamont is perched on the edge of an ancient floodwater pathway some eight miles wide.

Coupled with his text, the map is a test for the imagination. What Bretz reported, based on his field observations, is that large expanses of Palouse hills—some reaching 200 feet in height—were obliterated. The relatively few hills that remained within the scabland tracts exist as individual islands or small archipelagoes.

Bretz's gift as a writer surfaces throughout his scientific papers. His supple description of how the Spokane flood reshaped the surviving Palouse hills is one example. His purpose was forensic, yet he also captured the austere beauty in how the surviving "loessial" hills appear as remnants of an embattled fleet. The floodwaters had carved into and steepened their slopes, leaving high-blunted faces in the form of "great prows" generally facing Spokane, with streamlined tails pointing downstream toward Pasco.

"The nose of a prow may extend as a sharp ridge from the scabland to the very summit of the hill," he wrote. "It is impossible to study these prow-pointed loessial hills, surrounded by the scarred and channeled basalt scablands, without seeing in them the result of a powerful eroding agent which attacked them about their bases and most effectively from the scabland's up-gradient direction."

For the most part, the postcard images of the Washington scablands are from places eighty miles or so to the west of Lamont, where the floodwaters carved out huge cataracts at Dry Falls, Potholes Coulee, Frenchman Coulee, and Grand Coulee.

Yet the power of the floodwaters in the easternmost tract, up against the Palouse, was no less formidable. After leaving giant ripples on the plains near present-day Fairchild Air Force Base, the hammering waves blasted through a layer of Wanapum basalt near Four Lakes (twelve miles west of Spokane on Interstate 90), exposing the 1.4 billion-year-old, bluish-green sedimentary and metamorphic rocks of the Wallace Formation.

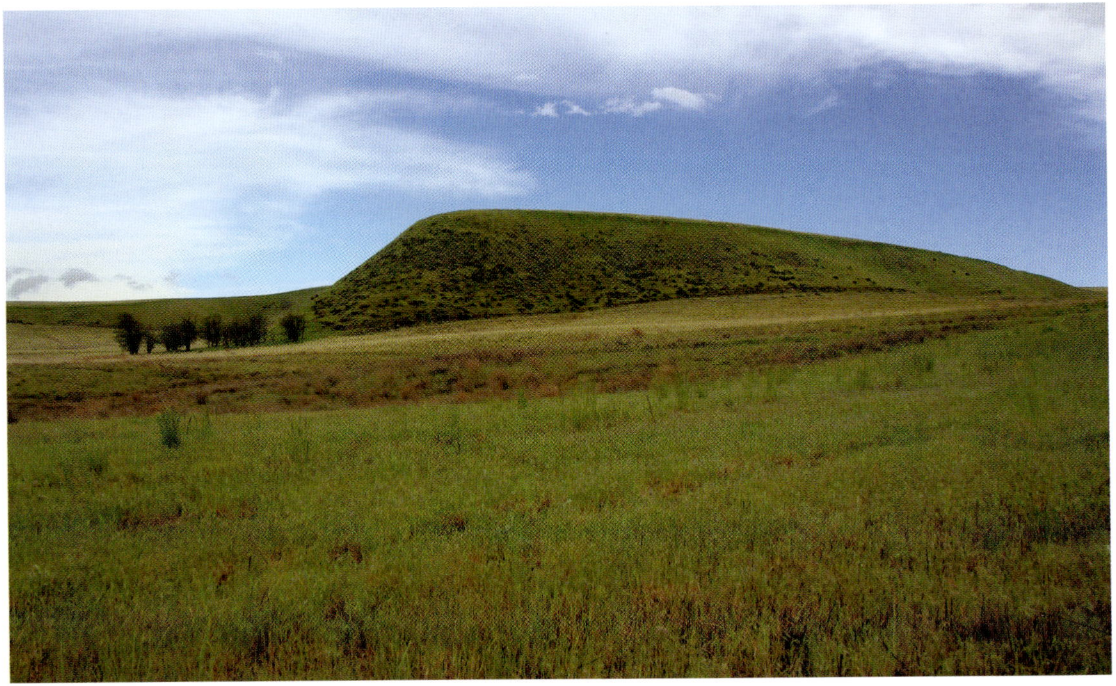

Bretz hill near Palouse Falls.

Farther south, near Lamont, the Lake Missoula floodwaters administered a gouge at Rock Lake that reaches nearly 400 feet below the lake's surface. In the surrounding Palouse hills there is evidence that the waters topped out at an elevation approaching 2,500 feet near Williams Lake, nearly 800 feet above the surface elevation.

The environmental and economic legacies of the flood pathways are in plain sight. Where Palouse hills survive, farmers plant wheat and lentils. Where the hills have been washed away, cattle graze. There are few lakes and wetlands in the Palouse, whereas the scabland tracts give rise to hundreds of them, including the lush maze of ponds, streams, and lakes that compose the Turnbull National Wildlife Refuge south of Cheney and the Columbia National Wildlife Refuge north of Othello. If you see mule deer in Washington State, chances are you are in or near the scablands. The same is true for the floral eruptions of native camas, Rocky Mountain iris, balsamroot, and green-banded mariposa lilies, among others.

Bretz offered layers of evidence for his Spokane flood theory and certainly, in his mind, each observation he framed defied other explanations. The haunting Bretz hills (he referred to them as "Palouse islands") are in this category, a signature of his science but also his aesthetic appreciation for a landscape that deeply intrigued and inspired him.

The road out of Lamont becomes McCall Road and dives farther south for ten miles before making a turn to the west. As it begins to bend westward, the road offers an open view for miles to the southwest. And there, in the distance, you can see a dreamlike fleet of streamlined Bretz hills, prows pointed north and slightly to the east. They look lonely and defiant, and seem to harbor a memory.

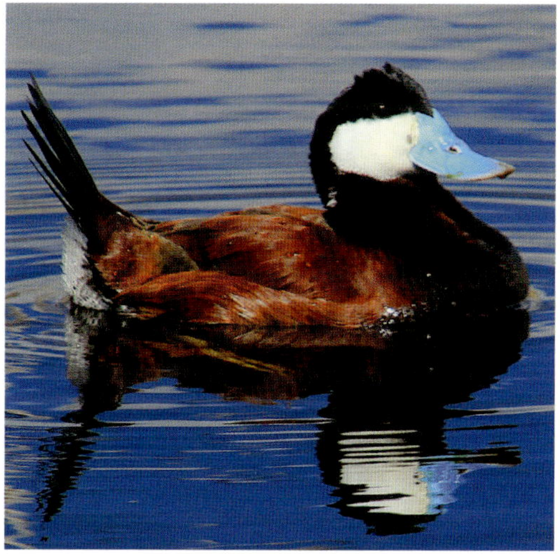

Ruddy duck on a scabland pond.

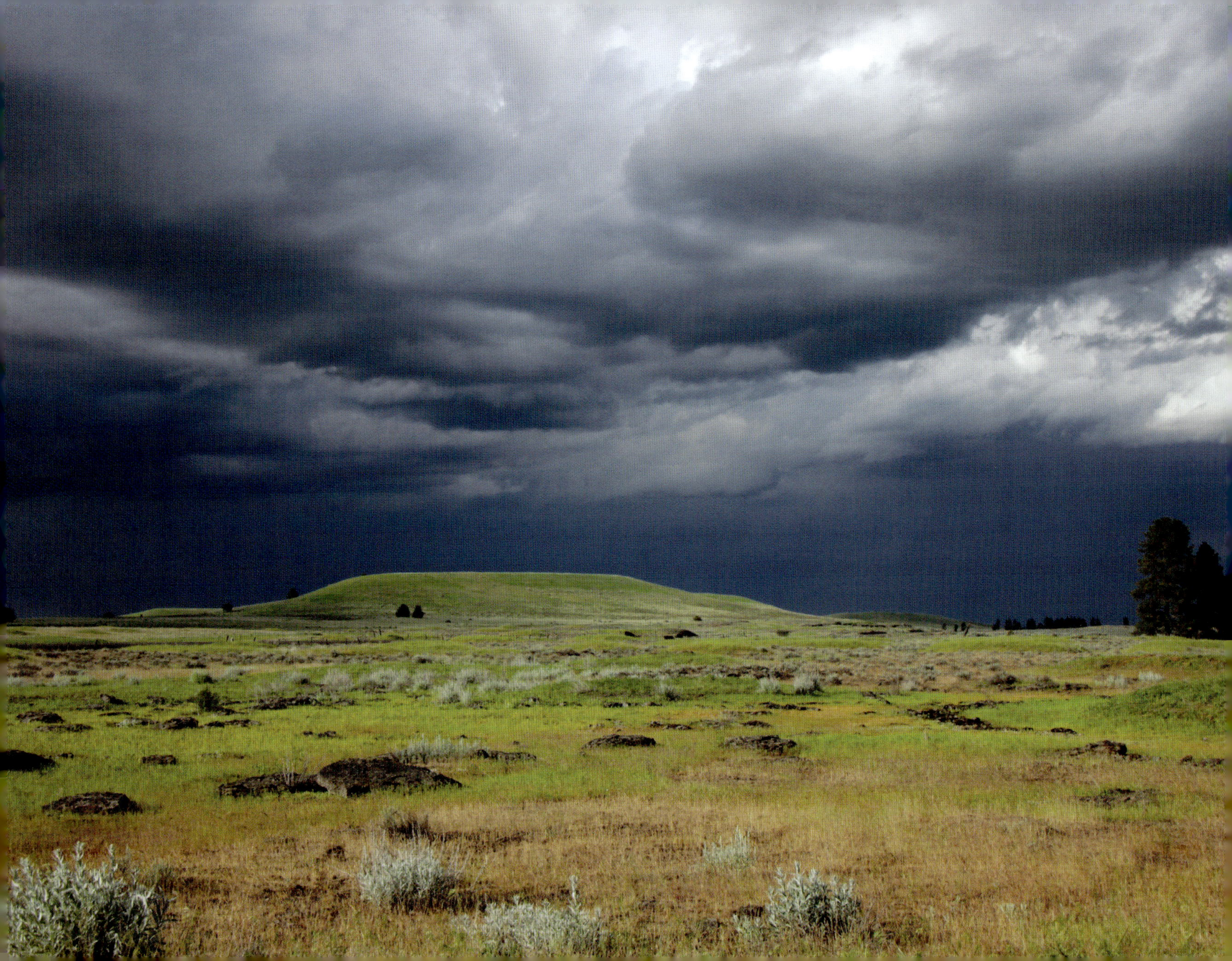

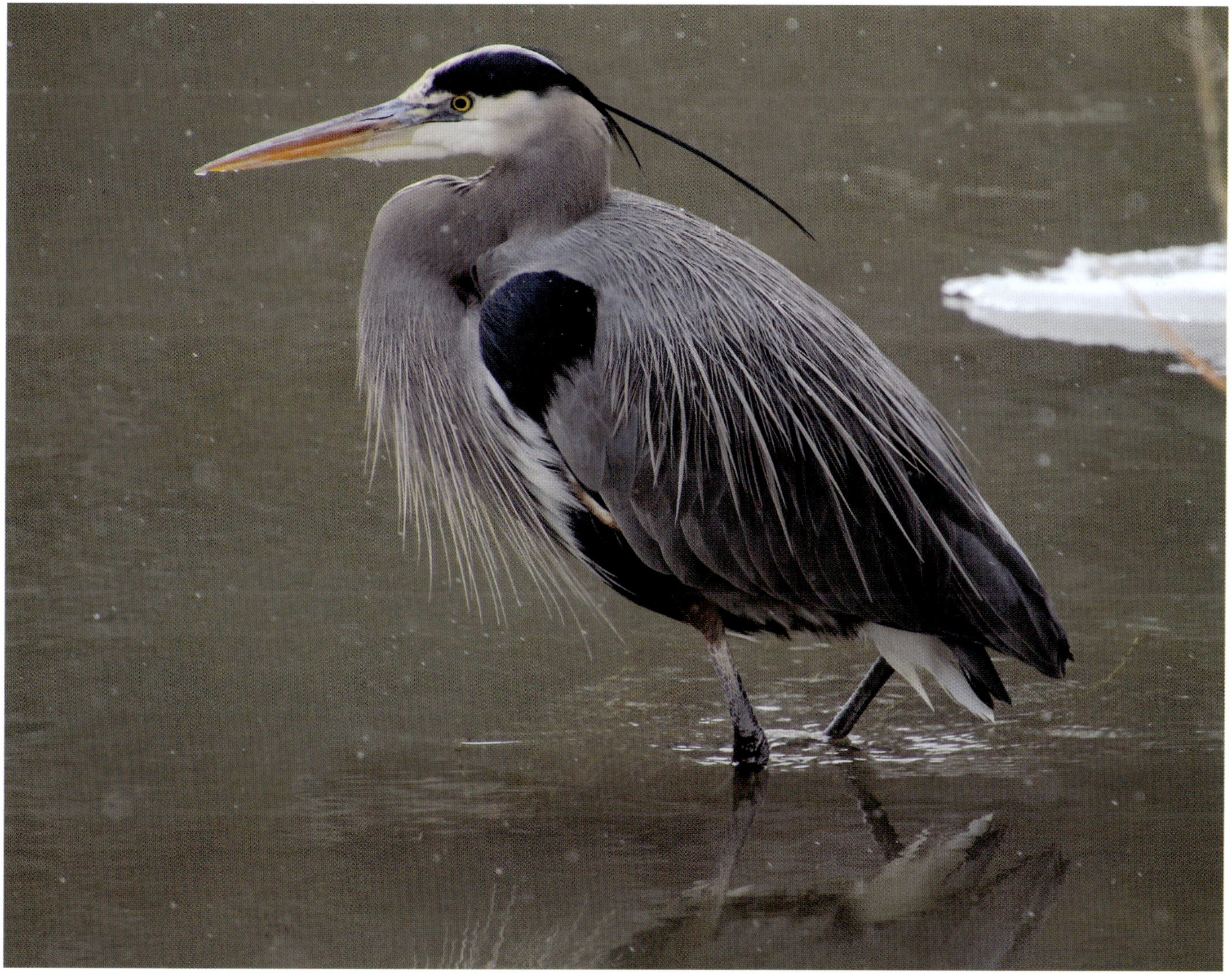

Left: Bretz hill and scabland terrain near Sprague. *Right:* Great blue heron on an icy creek near Spokane.

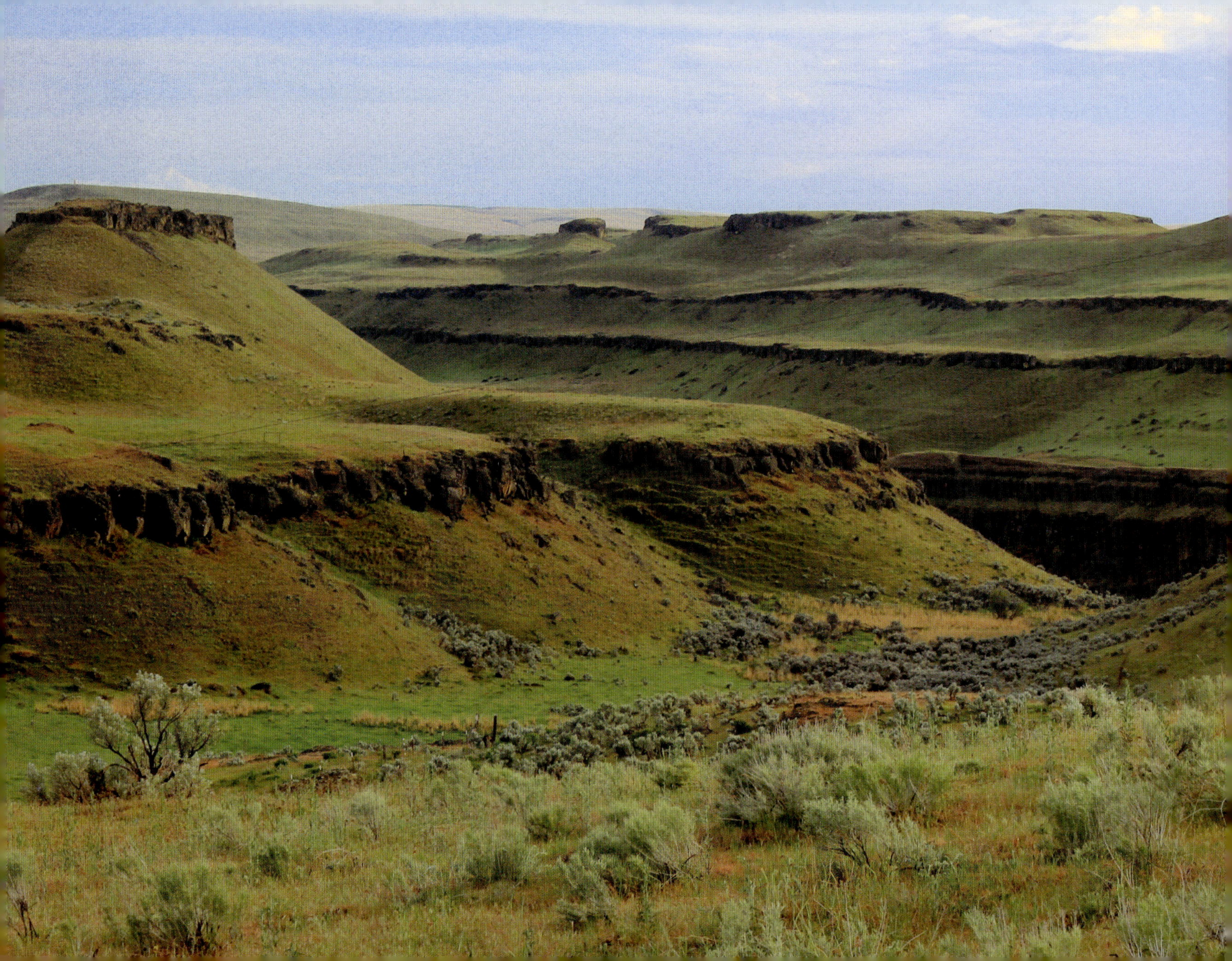

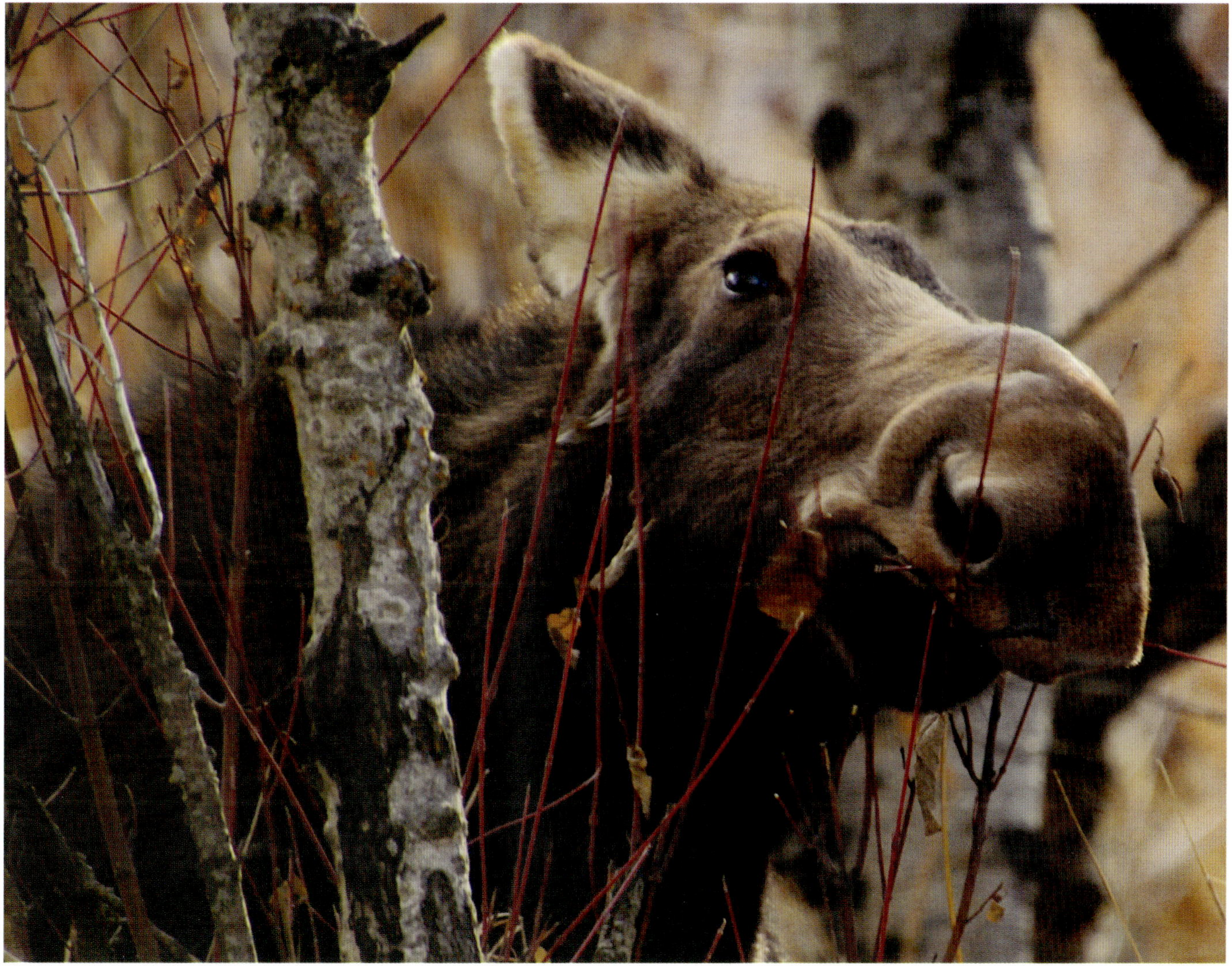

Left: HU Ranch Coulee east of Washtucna. *Right:* Moose munching on red-twig dogwood, Turnbull National Wildlife Refuge.

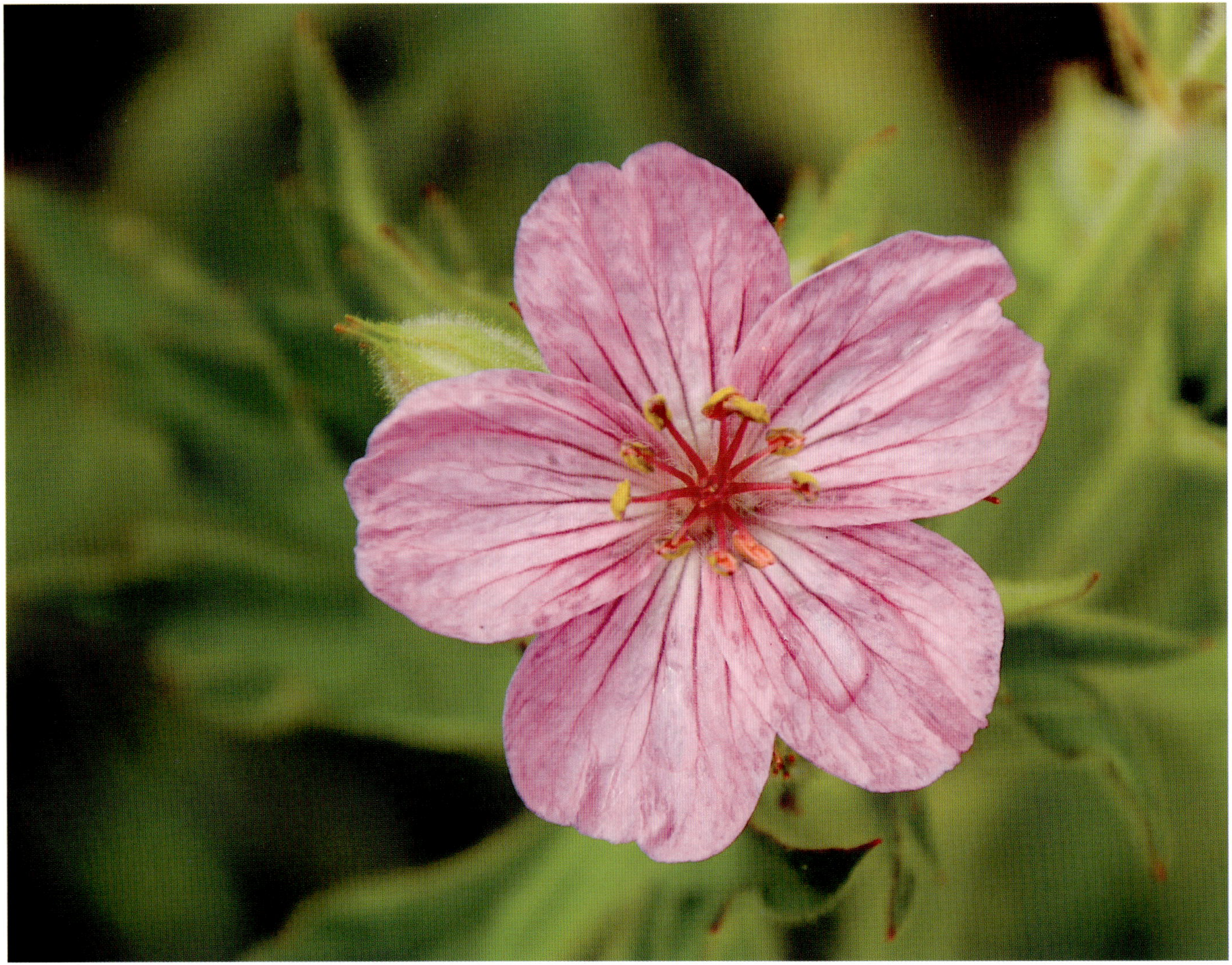

Left: Wild geranium. *Right:* Scabland terrain near Fishtrap.

prominent landmark in Whitman County because its 3,600-foot peak kept its nose above the floods of basalt. But the butte would sport a snow-streaked alpine peak were it not for the regional subsidence caused by the weight of the immense layers of cooled lava. A basic consequence of this sinking earth story is that a free drop of water in the inland Northwest is generally inclined to flow toward Pasco, and Rock Creek is a part of that hydrological web.

Three miles downstream from the spatter piles, the creek tumbles through the natural spillway at Towell Falls. From there it flows another five miles before offering its water to the main stem of the Palouse River. The meandering river then becomes the border between Adams and Whitman Counties for several miles. As it gathers Cow Creek at Hooper (pop. 54), the Palouse is making a gentle, westward arc toward Washtucna (pop. ~200) and its namesake coulee.

The strange thing is, the Palouse River doesn't make it to Washtucna Coulee. And because it doesn't get to Washtucna, it no longer flows by present-day Connell (pop. 5,566) and then on to Pasco, where it used to give its water to the Columbia. Instead, three miles east of Washtucna, the river turns on a dime. Like a point guard jabbing into a crossover dribble, the river makes a ninety-degree turn and heads straight down a natural, basalt-lined chute to Palouse Falls.

Just why the Palouse makes its pivot is something Bretz first explained in 1923. In spare prose, he wrote about how the "glacial flood from the north" poured into the Palouse River valley in several places along a thirty-mile stretch between present-day Washtucna and Winona. The valley was far too small to contain the epic floods, he explained, and "leaks" occurred where the water and ice overflowed.

Bretz's use of the word "leaks" to describe the overflow of the river valley was a bit like choosing the phrase "switching error" to describe a head-on train wreck. But his cautious word choice likely reflects his awareness that he was writing for an audience of his peers, most of whom would be skeptical and some of whom would bristle at the seemingly supernatural flood that was central to his theory.

In reality, though, what he was describing as "leaks" were events more powerful than hydrogen bomb explosions—overwhelming pulses of water, rock, and ice blasting through uplands that had previously formed the divide between the Palouse valley and the Snake River gorge. Once the divide was breached, gravity took over and the Palouse River naturally found the new, steeper path to the Columbia River (via the Snake). In doing so, it shifted its course some thirty miles to the east.

The result, today, is a picturesque illusion. It's the extensively photographed sight of the Palouse River swerving around bladelike pinnacles of basalt, falling 185 feet into an enormous plunge pool, and then flowing into one of North America's most beautiful gorges. It may say Palouse River canyon on the postcards and maps, but the river had nothing to do with its namesake canyon. It begs for an asterisk of clarification and the hope that someday a poet will write wistfully and well about Pasco's missing river.

Right: Palouse Falls.

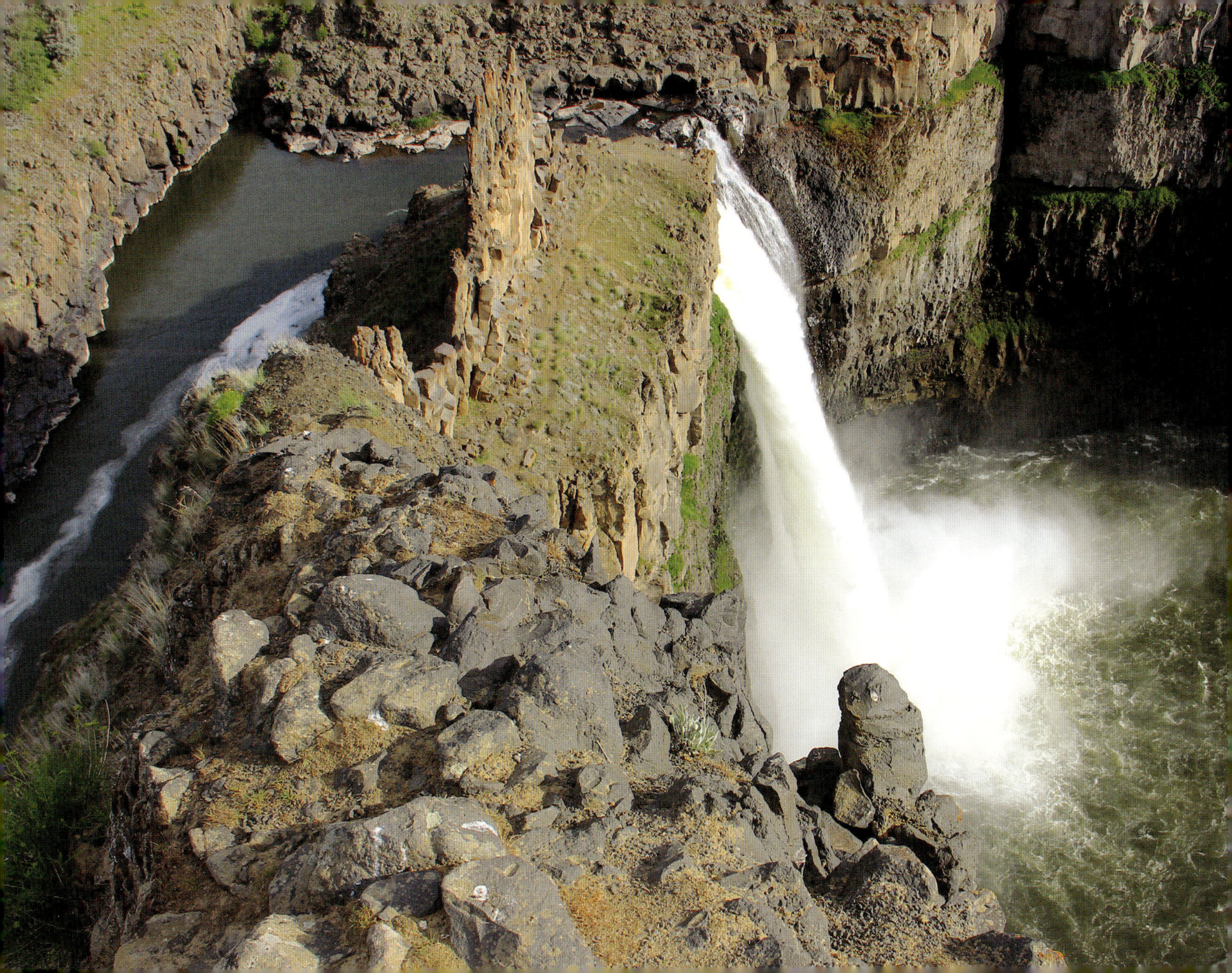

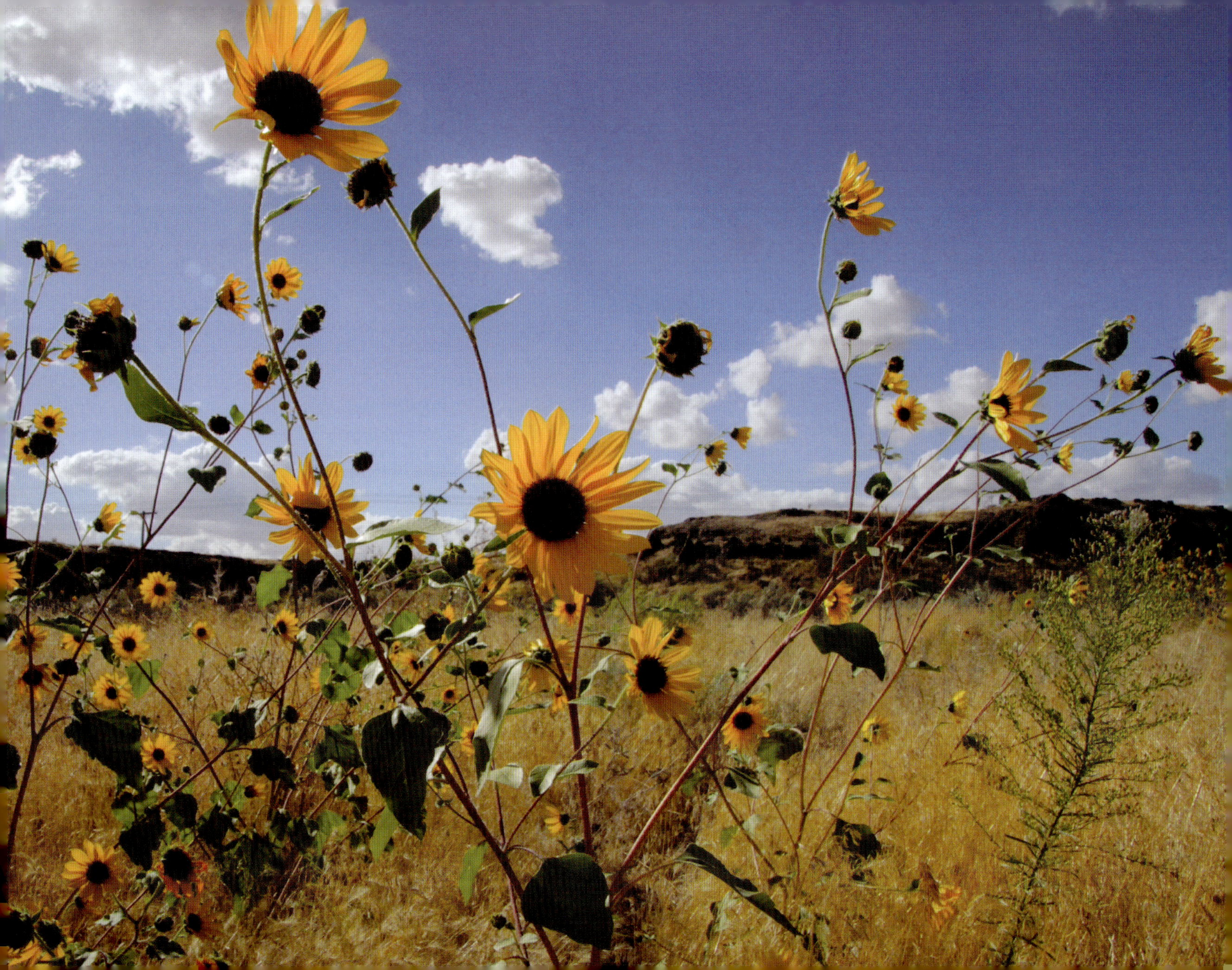

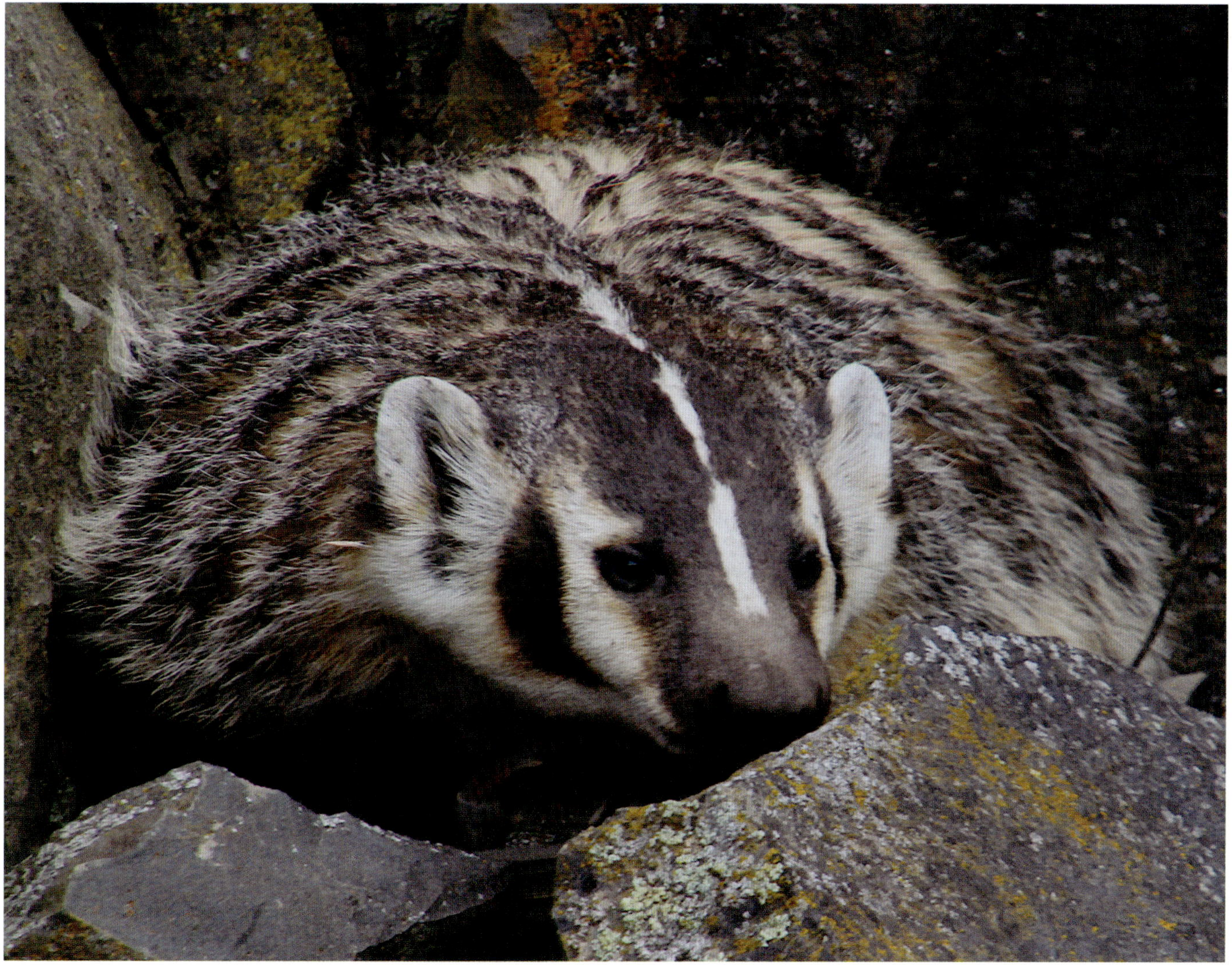

Left: Wild sunflowers, Palouse River canyon. *Right:* American badger.

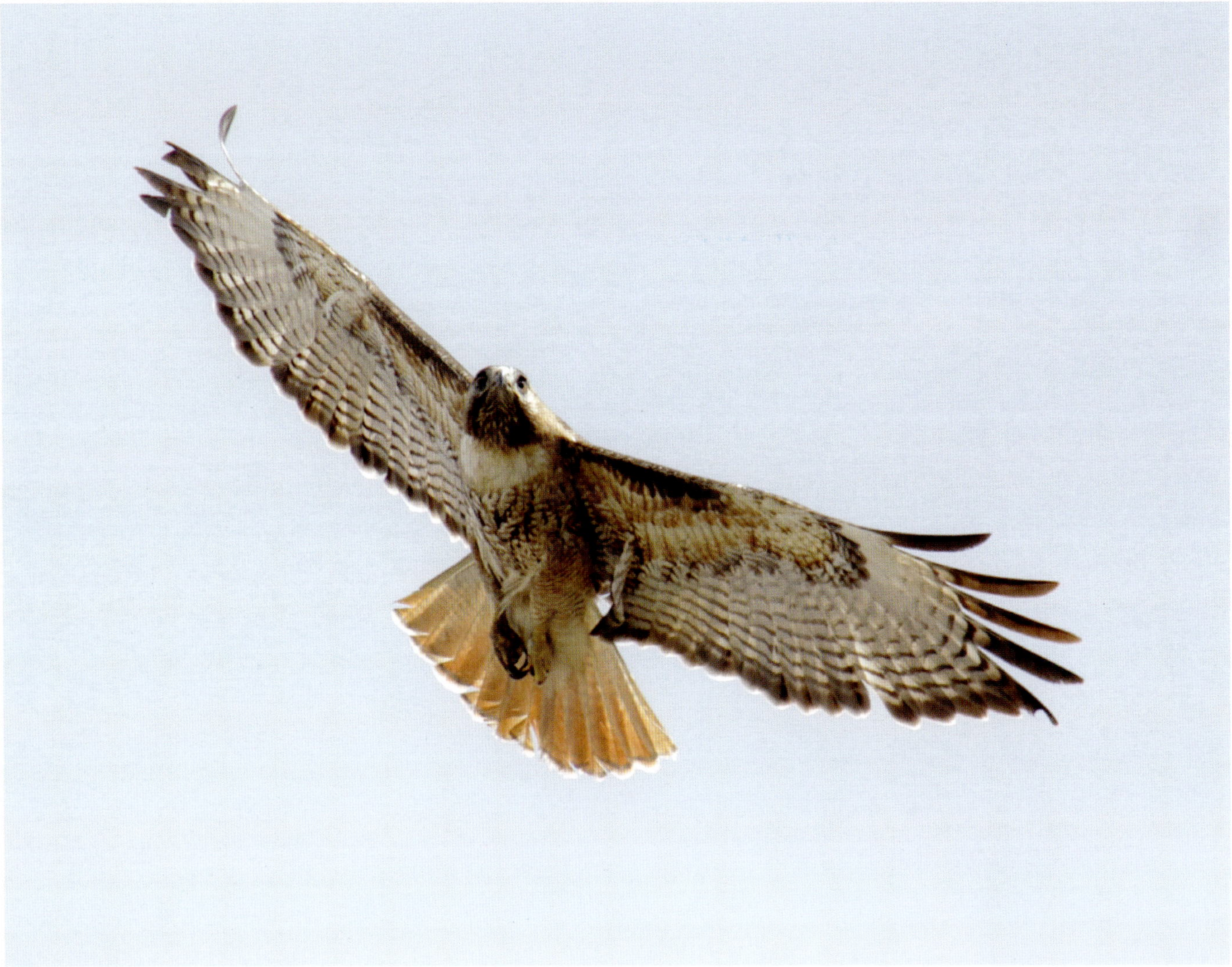

Left: Red-tailed hawk, Adams County. *Right:* Rock Creek valley, Whitman County.

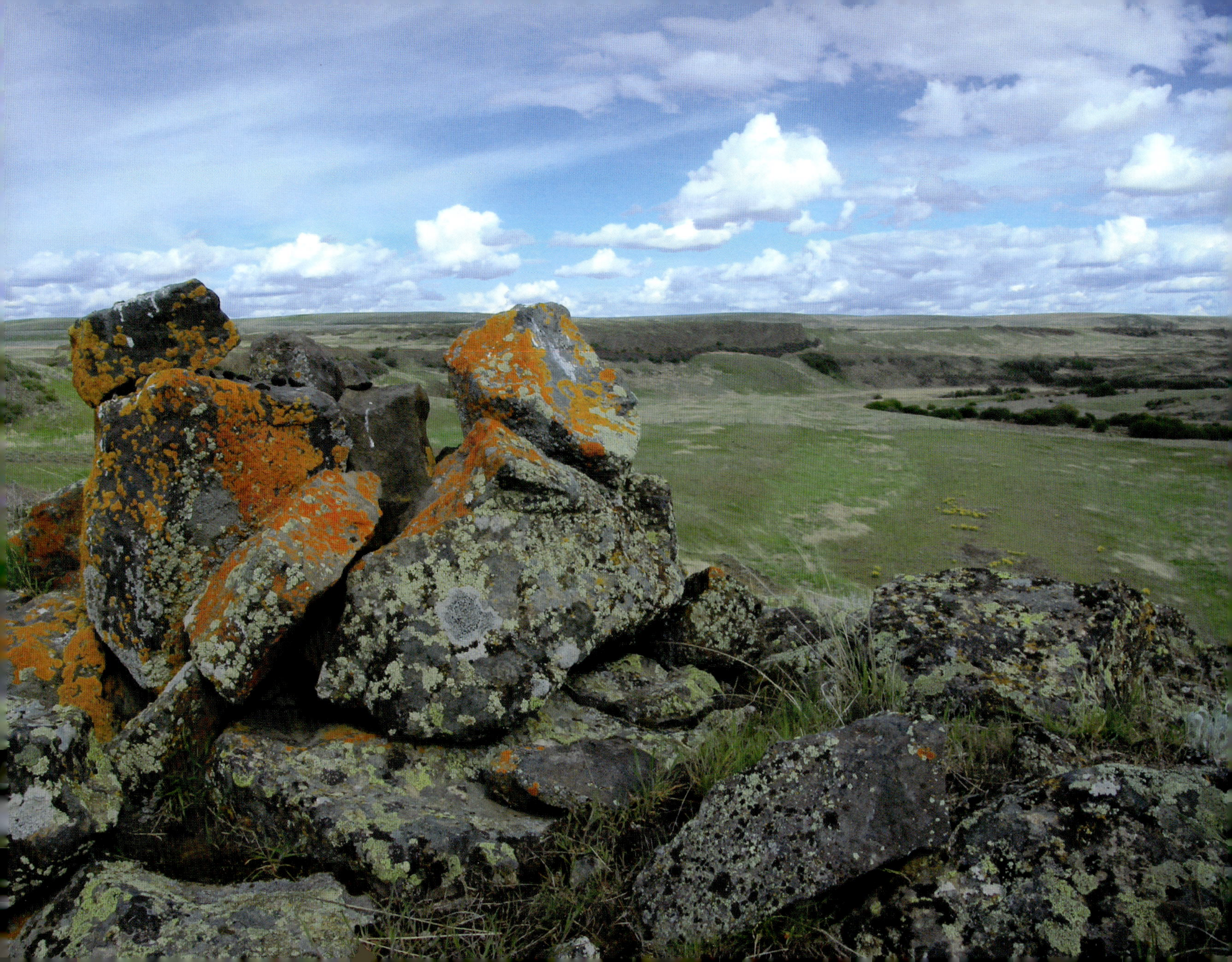

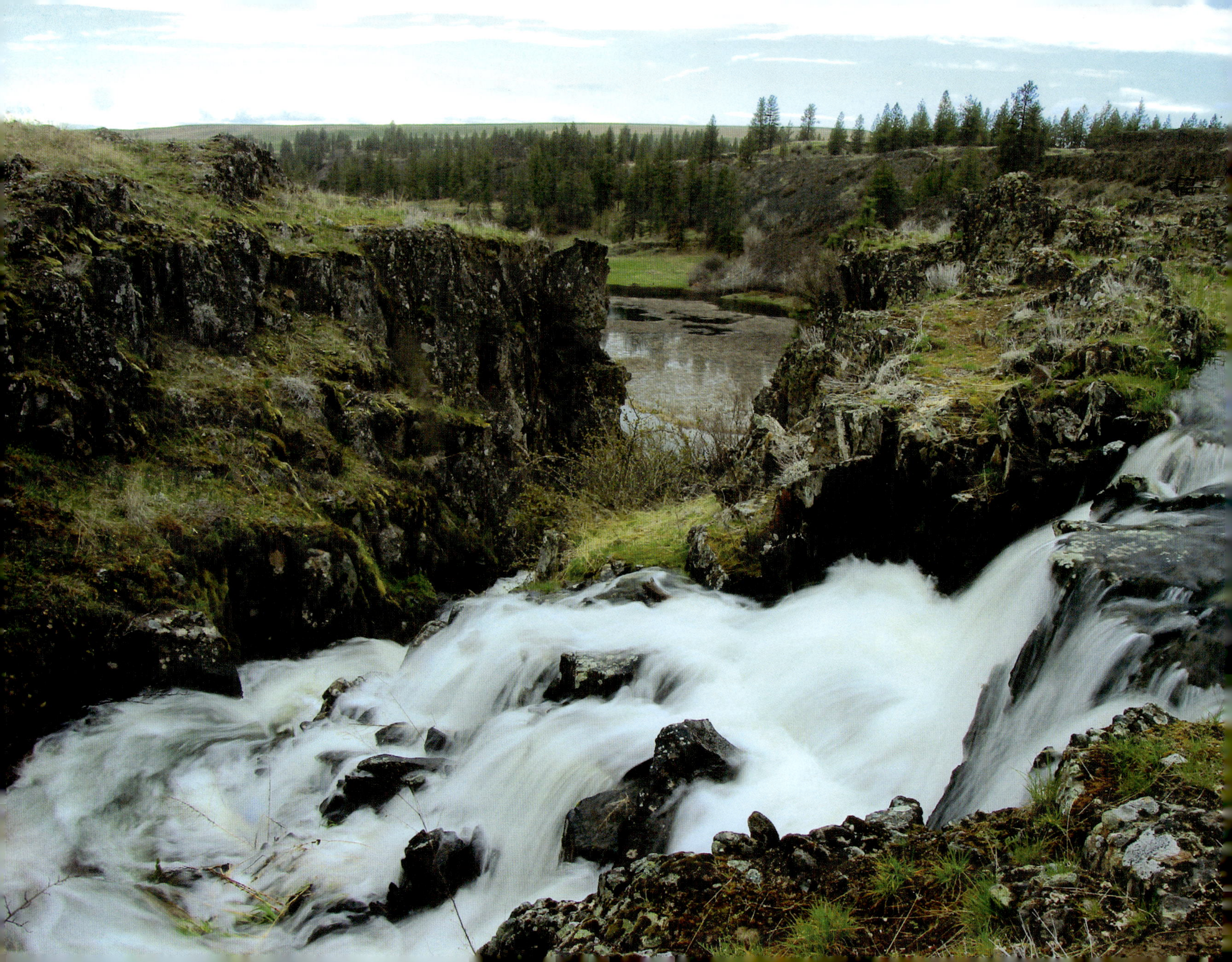

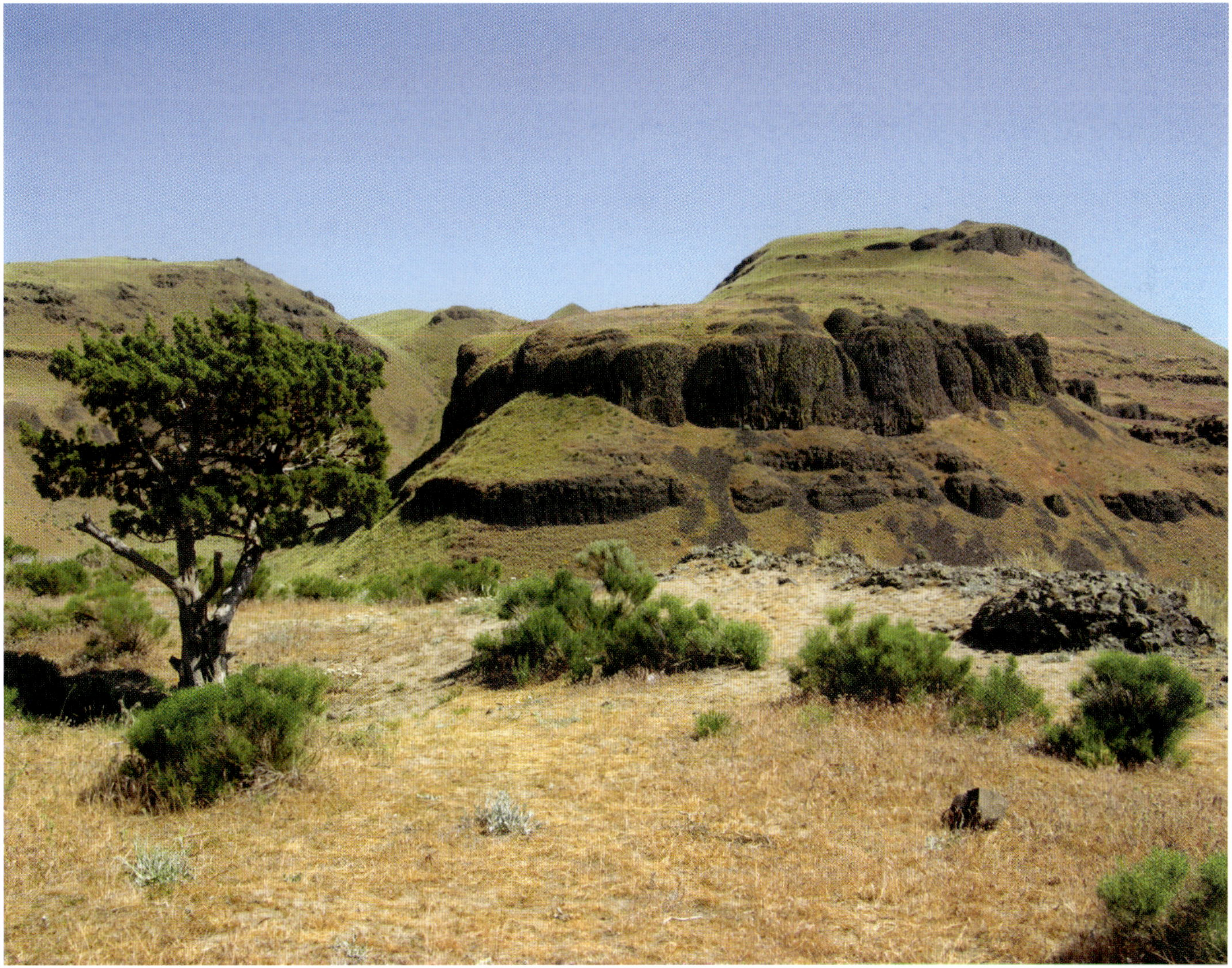

Left: Spring runoff near Badger Lake. *Right:* Lone juniper at Wallula Gap.

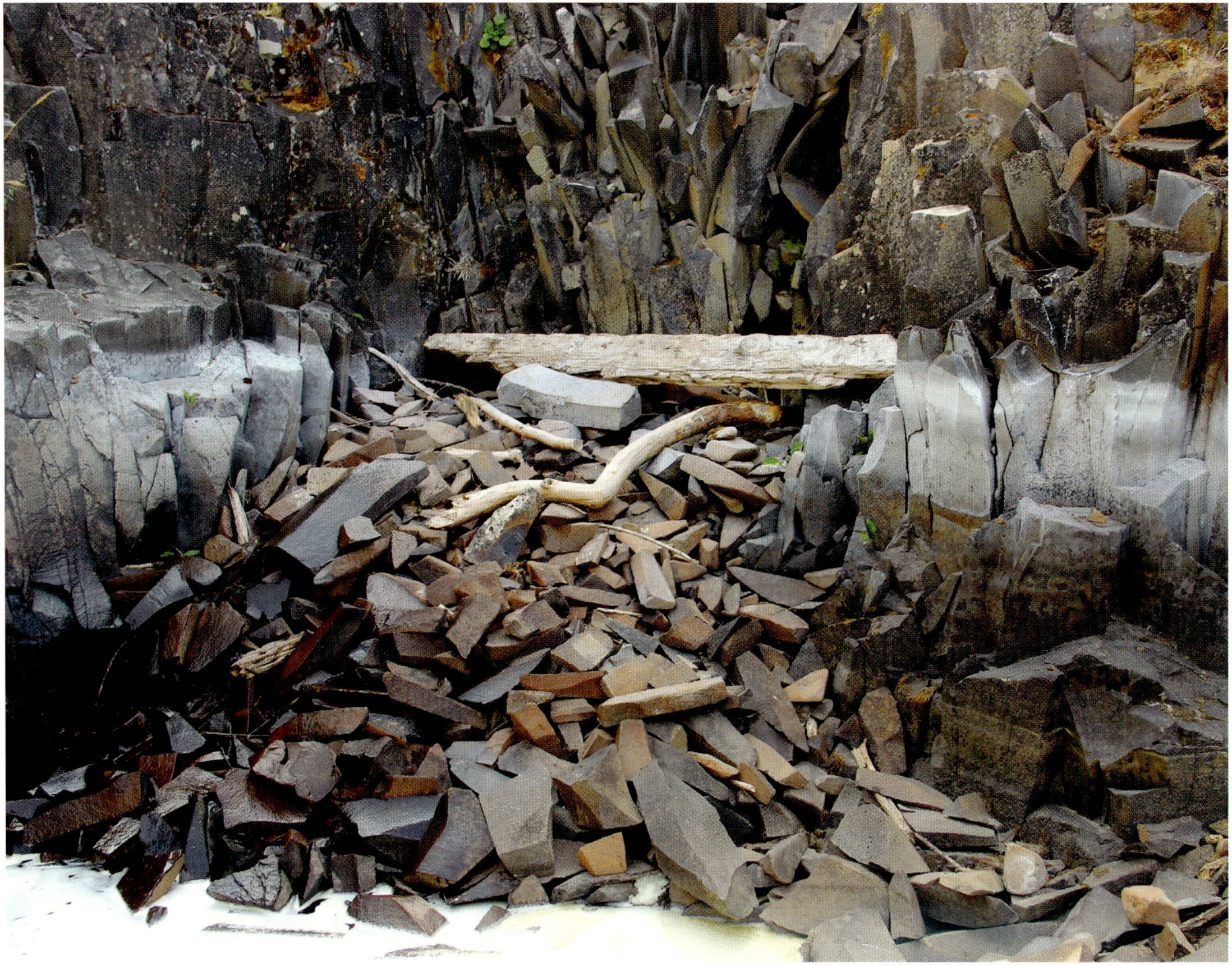

Left: Basalt talus near Fishtrap. *Right:* Migrating tundra swans near Sprague.

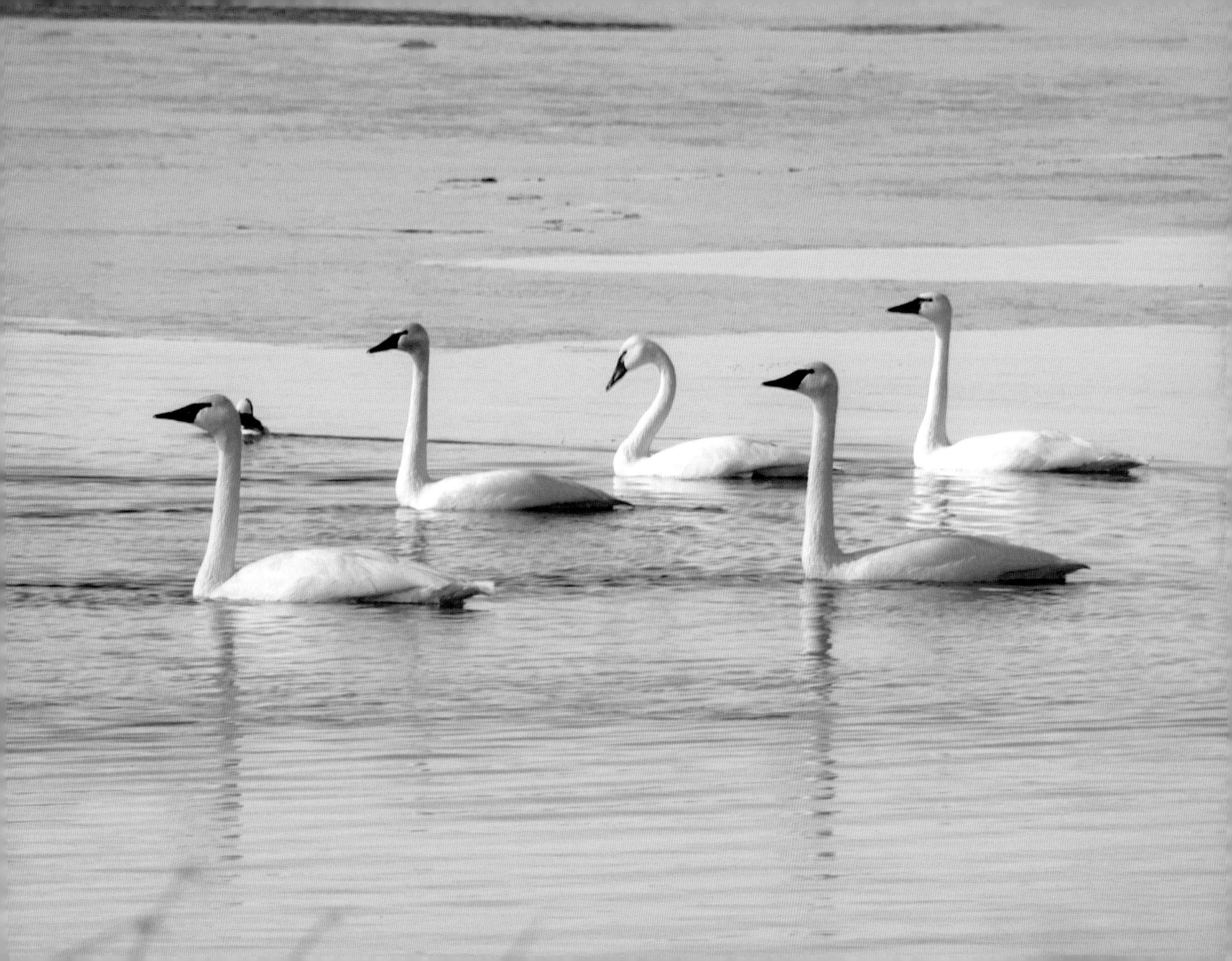

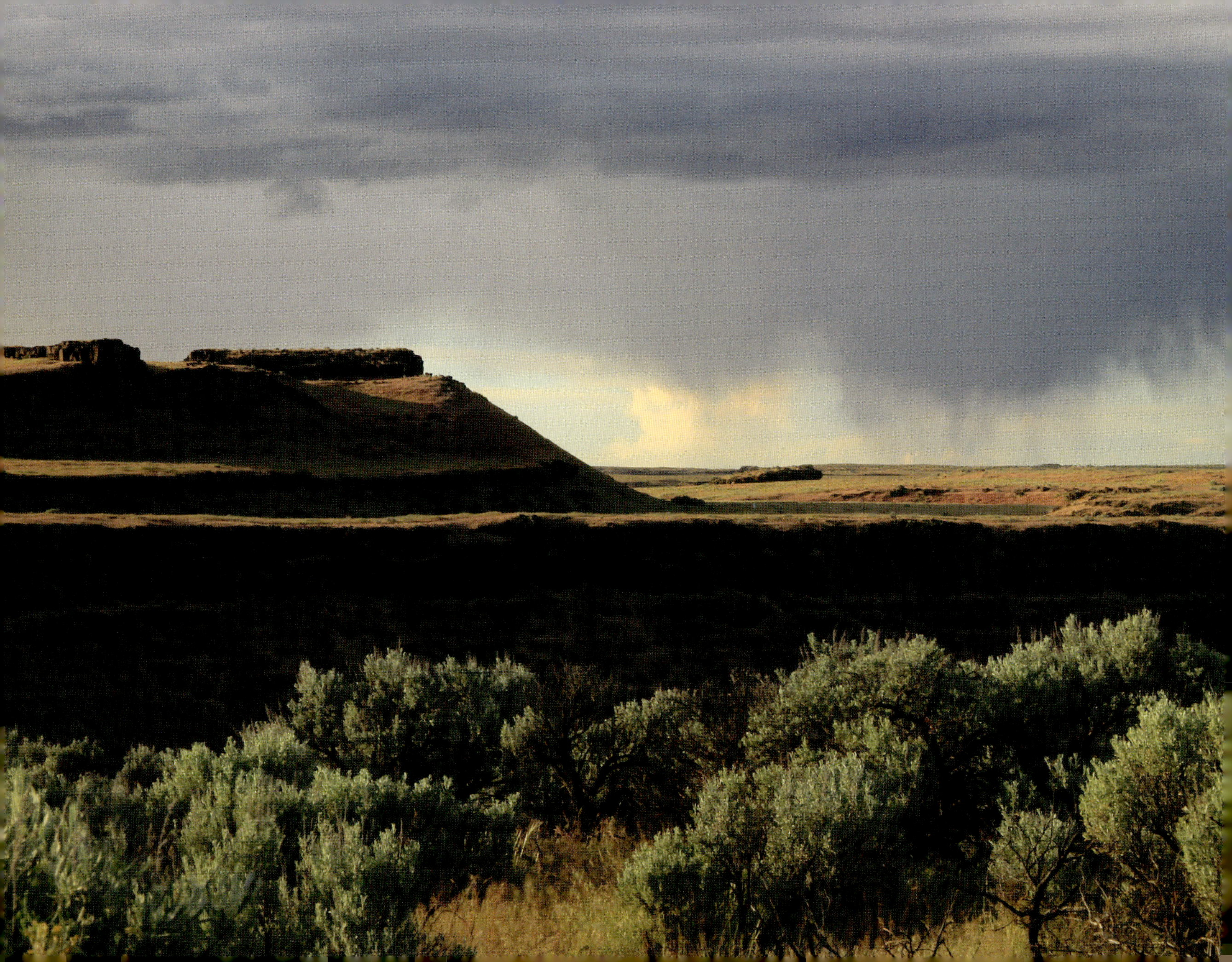

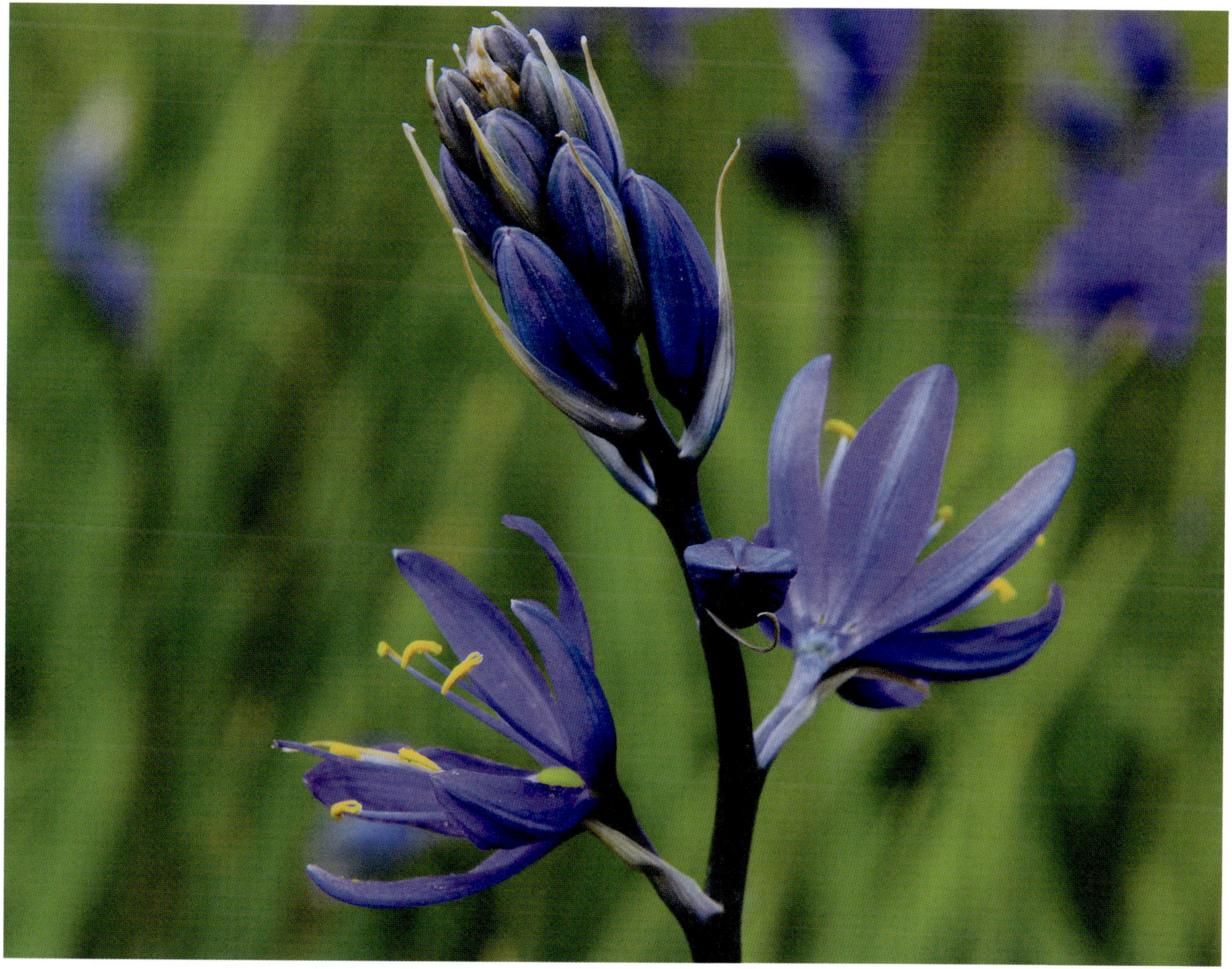

Left: Spring storm south of Moses Lake. *Right:* Blooming camas.

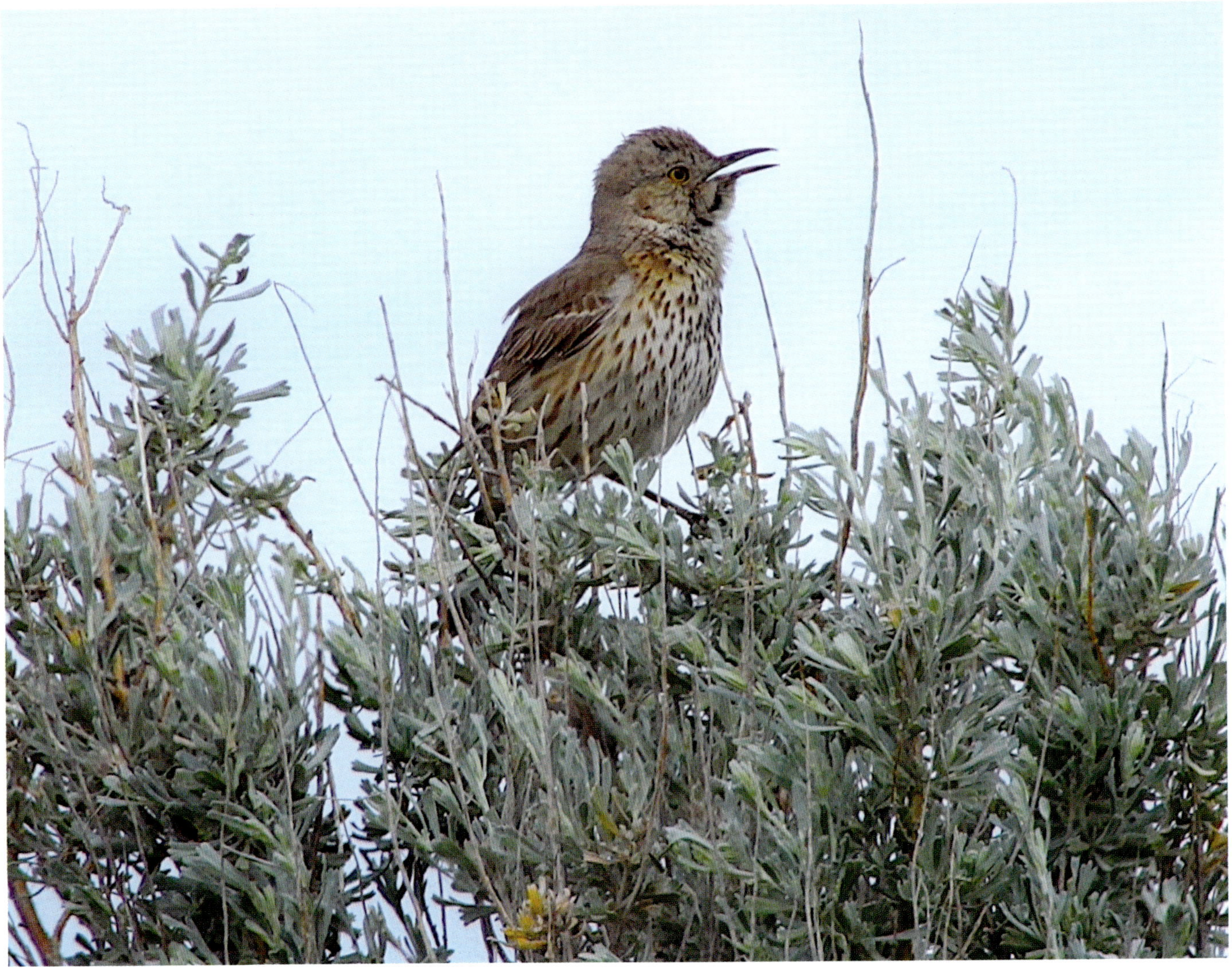

Left: Sage thrasher north of Odessa. *Right:* Lichens on Grande Ronde basalt, Spokane River gorge.

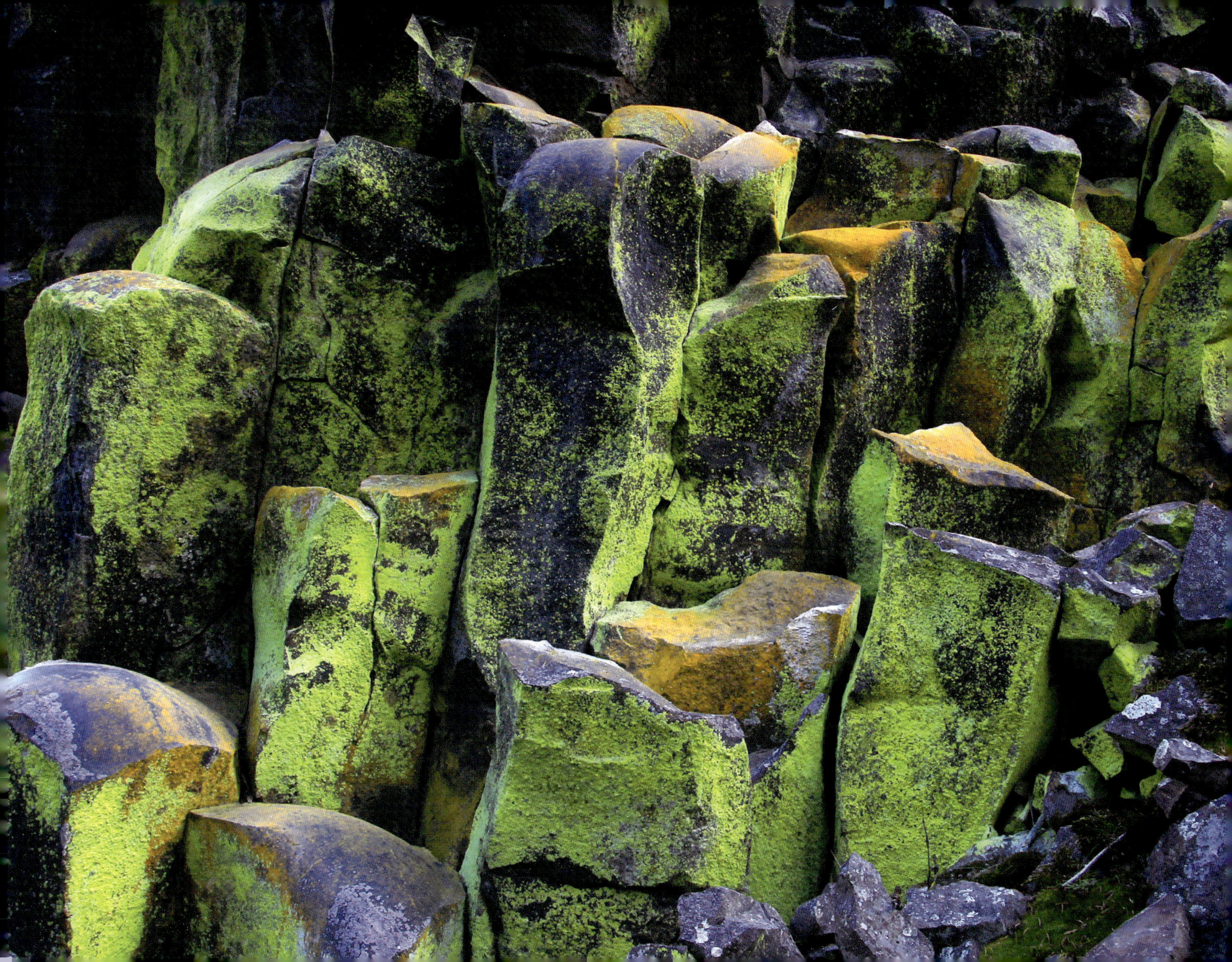

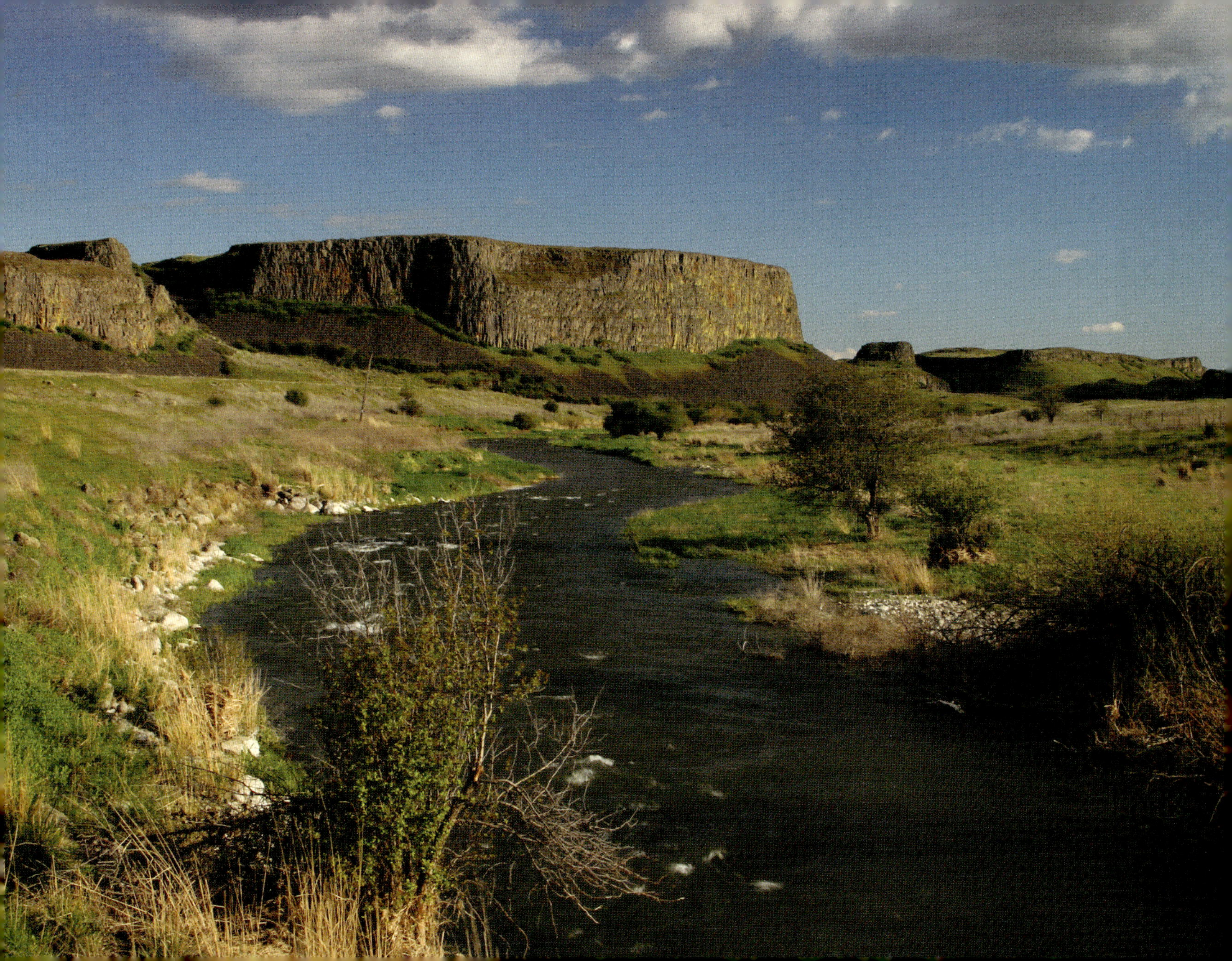

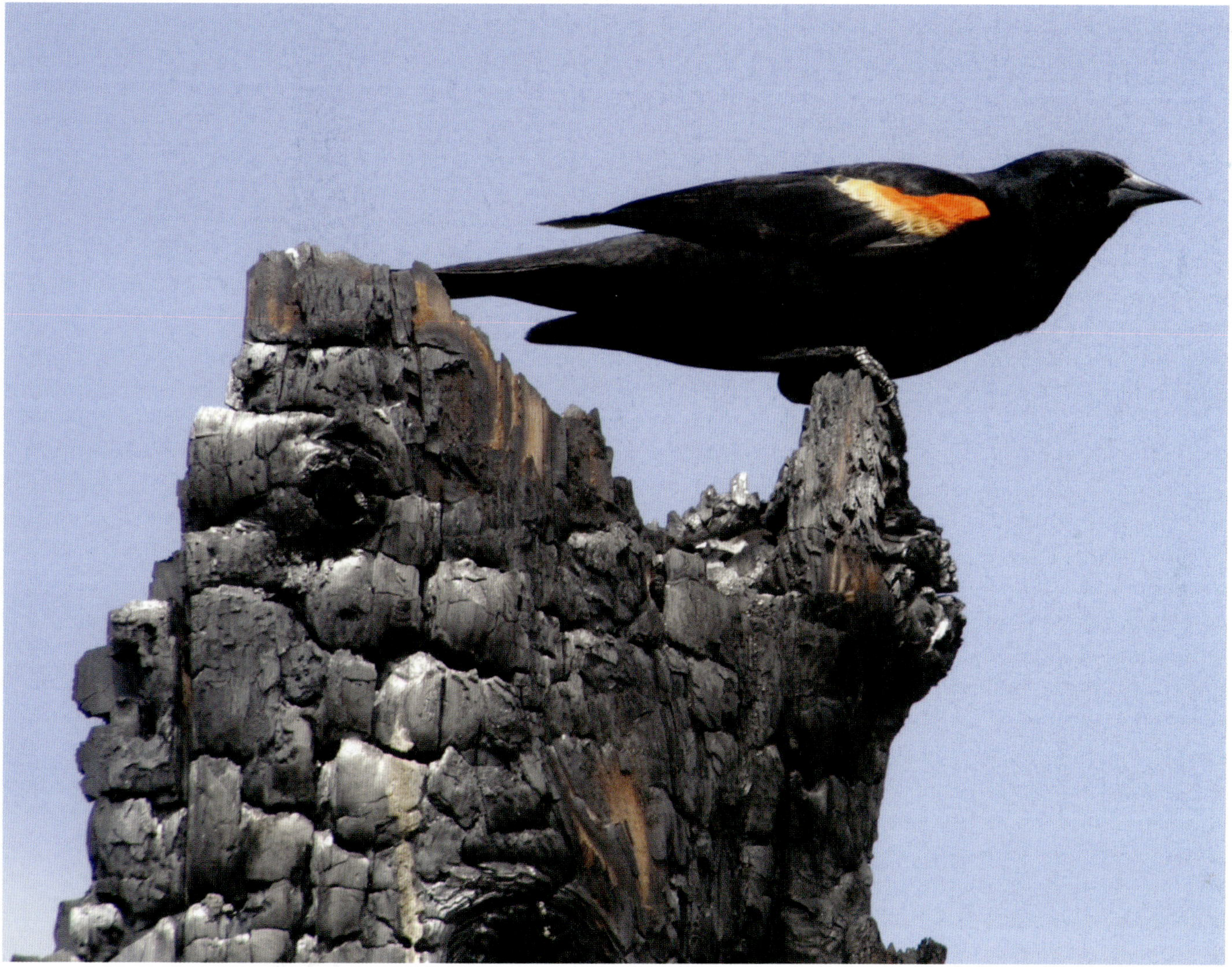

Left: Rock Creek at Escure Ranch. *Right:* Red-winged blackbird.

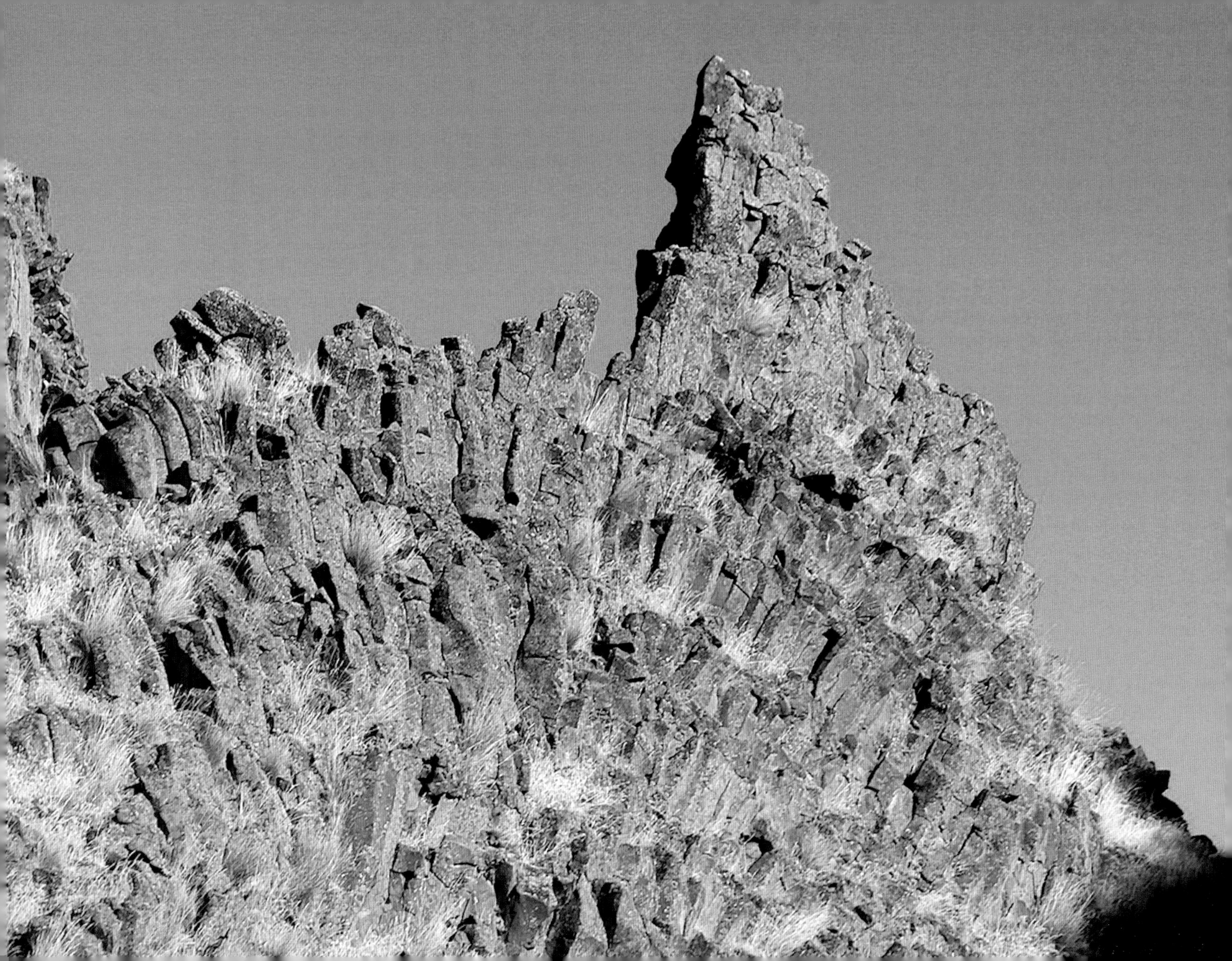

CHAPTER VI.

An Angel's Wing in Devil's Canyon

From her brothers I have stories of my mother as an athletic teenager, yet I have no memory of her being able to bend her knees. In her early twenties she was stricken with severe rheumatoid arthritis that maimed her joints. To adjust for this disability, Gil built a long wooden ramp to the landing by the side door so she wouldn't have to struggle up the steep concrete steps at the Hartman home in east Pasco. She and my older sister were living there with my grandparents in December 1956 when it was time for me to be born. Because my father was on active duty in Korea, I was eligible to be delivered, at government expense, by an Army doctor. Gil just had to drive us twenty miles to the west, to the military field hospital at the Hanford Nuclear Reservation.

A quarter century later I began writing about the lesser-known, grim legacy of my birthplace. During the Second World War, the 560-square-mile site had been cordoned off as part of the top secret Manhattan Project. The towns of Hanford and White Bluffs were wiped from the map, with well over 100 caskets exhumed and reburied in Prosser, another small town just to the south of the restricted zone. Inside the new fences and barbed wire, physicists, engineers, and thousands of construction workers hastily built the world's first

Left: A "wing" of Pomona basalt columns flaring in Devil's Canyon.

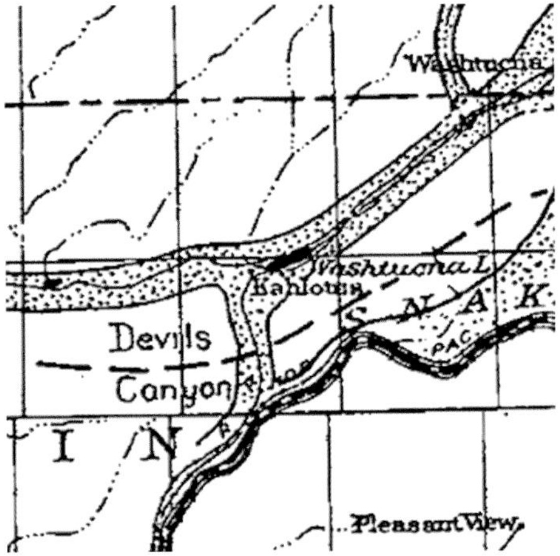

Devil's Canyon in Bretz's 1923 map of the scablands.

plutonium factories. A small sphere of Hanford plutonium was at the core of the world's first nuclear explosion in New Mexico in July 1945 (the so-called Trinity test), and another was in the "Fat Man" atomic bomb that a B-29 dropped on Nagasaki, Japan, less than a month later.

Japan surrendered. Hanford kept going. As the Cold War ramped up in the late forties and fifties, several more graphite-block reactors were built along the banks of the Columbia River. More battleship-sized plutonium refineries were added to a sage-wrapped plateau below Rattlesnake Mountain. Because the plants were deemed vital for national defense, they were exempt from federal and state environmental laws for a half century.

I began making visits to Hanford as a journalist in 1981, writing for regional magazines and newspapers. One chain of unanswered questions led to another, and in 1985 I was hired by a new citizen organization to lead a more systematic effort to unlock Hanford's secrets. Along the way I rediscovered the work of J Harlen Bretz who, long before the Manhattan Project, had visited the area as he was tracking the course of the ice age floods.

"Like roads to Rome, all scabland rivers led to the Pasco Basin," Bretz had written, based on his fieldwork in the 1920s. And when the glacial floodwaters swirled into the Pasco Basin—which encompasses all of what is now the Hanford site—they dropped huge loads of cobbles and gravel, hundreds of feet deep. One of the largest piles is known as Cold Creek bar. It was atop Cold Creek bar that the largest of Hanford's secretive plutonium refineries were built.

This was purposeful. The so-called "Pasco gravels" at Cold Creek bar served as an enormous drain for the vast quantities of liquid waste generated by Hanford's atomic sluiceways. You could dump radioactive and chemically tainted effluent into the ground. The contaminated water would quickly percolate away, only to emerge years later in shoreline springs along the Columbia River. These and other revelations became part of a large body of troubling findings that, by 1990, led to the shutdown of Hanford plutonium production and a landmark cleanup agreement to try to contain the site's lasting and mind-boggling environmental contamination.

The Hanford assignment and efforts to reform and clean up the site consumed my life for several years. Unbeknownst to the Hartmans, Hanford had come right into their kitchen through their milk and water. Like the other good people of Pasco, they weren't told about the highly radioactive effluent dumped into the Columbia River just a few miles upstream from the Pasco drinking water intake. Nor were they told about the highly radioactive clouds of Iodine-131 discharged into the air by the early plutonium extraction plants. Those emissions—which contaminated forage (and thus, local milk supplies) throughout the lower Columbia basin—were especially dangerous for infants and children (especially females) because it put them at greater risk for thyroid cancer and hypothyroidism. My mother was eleven the first time she was unwittingly exposed to Hanford radioiodine. Years later she was diagnosed with hypothyroidism.

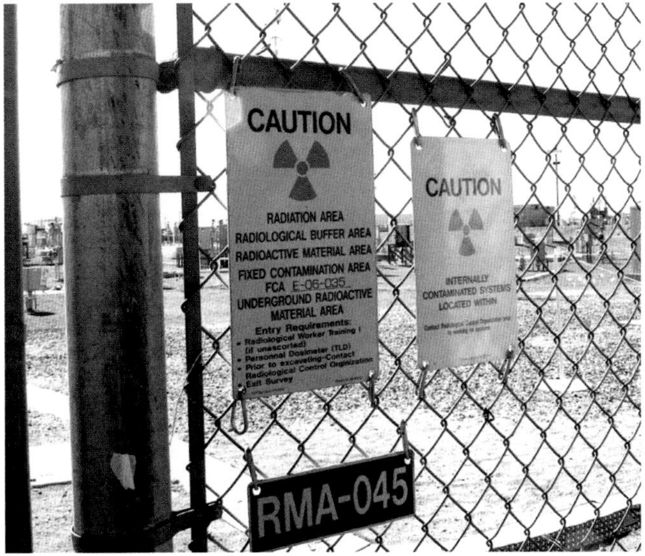

Hanford radiation warning signs at Cold Creek bar.

My work drove me deeply into the forensics of the Hanford saga, and I was fortunate to have the support and help of some of the nation's best scientists and medical experts. But it was the broader and lasting questions that weighed on my spirits. It's just not possible to disconnect Hanford from the moral turmoil of the brutal war that led to its creation and the subsequent stockpiling of what is arguably the world's most lethal substance. The Ozzie and Harriet–like simplicity of the Hartman home in Pasco—with its white picket fence, hedges of lilacs, and picture-window view of the Horse Heaven Hills—seems like it belongs in a more sanguine and less complicated world. But, of course, the world is prone to complications that can leap right over hedges and knee-high fences.

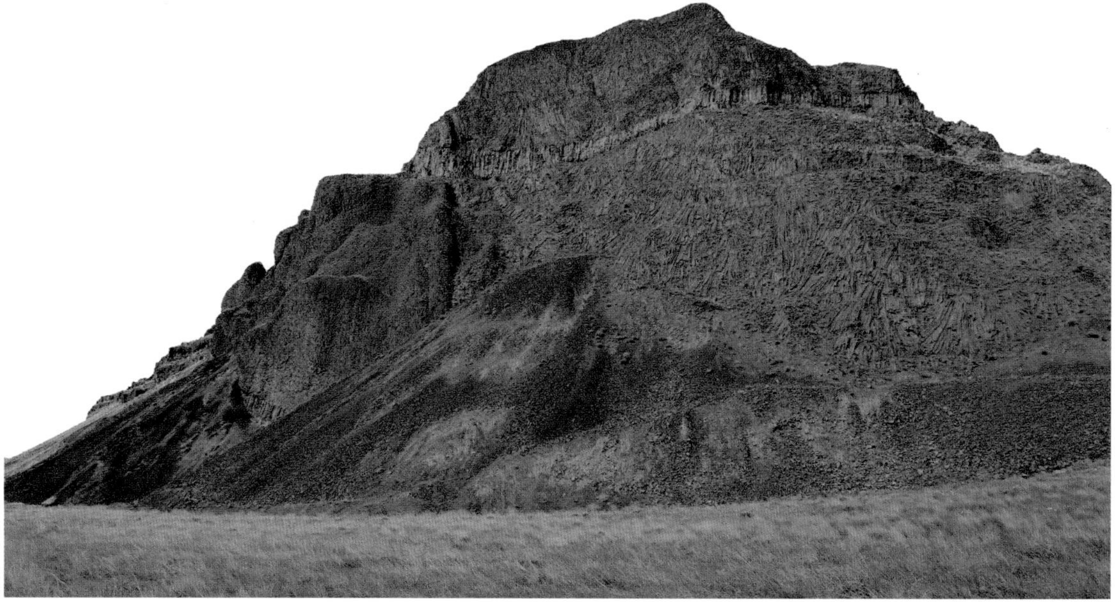

Devil's Monument.

For one of my early magazine assignments, I interviewed a small group of brave Hanford scientists who were choosing to speak out against the reactivation of the site's largest plutonium plant. I met them—at their request—in the basement of a Methodist church in Pasco. It was the same church where my great-grandfather had once served as pastor and where we had gathered, just four years earlier, to mourn Gil's passing.

The article about the dissident scientists became a cover story for the Portland

Oregonian's Sunday magazine, and it propelled my career. Twenty-five years later, I had married, fathered two children, and granted myself permission to move on to other stories and causes. New awards arrived and then a new book project. Life got better, then steadily worse. In a matter of months, I lost my marriage, a rewarding job, and two dear friends—one to cancer, another to heartbreak. Other tragedies ensued.

This is why I walked into the scablands with my camera.

I went there for solitude—to grieve and try to regain my footing. The photography followed because it was therapeutic. It became the means to reengage my imagination and to bring home images that offered me consolation and that certain quality of hope that arises when we find beauty in unexpected places.

During one of my most difficult months, I was drawn to Devil's Canyon just south of Kahlotus, about forty miles northeast of Pasco. Kahlotus (pop. 200) is platted within Washtucna Coulee and is home to a now-dry lake that Gil used to frequent back when the lake held water and a healthy population of bass. The canyon is less than a half mile from where the lake used to be. It is a deep and dry rift, reliably hot by midsummer, and devoid of people.

The canyon formed because Washtucna Coulee wasn't wide or deep enough to contain its pulverizing load of Lake Missoula floodwaters. In a geologic heartbeat the overflow jumped the rim at a near ninety-degree angle and behaved as a hydrogeological chainsaw, chewing a near perfectly straight, 500-foot-deep and five-mile-long gouge in the earth. This arrow of a gash reaches from Kahlotus to where the spillover floodwaters emptied into the Snake River, just above where Lower Monumental Dam is today.

Bretz was fascinated by Devil's Canyon and highlighted it, along with the capture of the Palouse River, in his provocative 1923 article in the *Journal of Geology*. It was one of his polite ways of saying to his skeptics: "Explain this."

The rocks in Devil's Canyon stack one drama atop another. The most recent chapter is the one Bretz highlighted—the sudden excavation of a riverless, dry canyon with all the ballistic evidence of an overpowering flood. But there's an older story that is at least as dramatic and still visible on the landscape.

It turns out that millions of years before the Lake Missoula floods overflowed Washtucna

Coulee, an ancient river flowed through the area. The river valley became a pathway for some of the younger Columbia River basalt flows—the Esquatzel, Pomona, and Elephant Mountain. What it really means is that black-orange rivers of glowing magma followed gravity into the ancient river drainage, vaporizing the water and igniting all sorts of biomass along the way. Ultimately, the three younger flows filled the entire river channel.

This was 11 million or so years ago. When the ice age floods that excavated Devil's Canyon arrived some 17,000 years ago, the younger basalt flows were more resistant to the onslaught than were the older flows, namely those of the Wanapum formation that compose the walls and bedrock of the canyon.

The result is a peculiar and beautiful feature. It's a butte, dubbed Devil's Monument, situated about two-thirds of the way down the canyon—two miles north of the Snake River. Although it appears as a monolith, it's actually composed of three distinct portions—each a remnant of the three younger basalt flows.

As they did at Hat Rock butte near Wallula Gap, the Pomona basalt cooled into narrow columns, about the thickness of baseball bats. As I was sulking and walking and climbing on an afternoon in late July, I would stop to admire and photograph the colonnade. Then I would resume my perspiring climb up the rocks, taking care not to twist an ankle on the talus. As I turned a corner toward the ancient river channel, I looked up and was stunned to see that a flood-battered wall of Pomona columns had been carved into a wing-shaped outcrop.

I had to laugh at the obvious irony and symbolism. An angel's wing in Devil's Canyon. There wasn't a rainbow. Doves didn't fly. But it still seemed like a sign from God in a place that others might dismiss as godforsaken. The temperature was near a hundred degrees. Crows were soaring and calling in a thermal above the surrounding cliffs. And I was grateful to have made it through another day.

Right: Recovering wildfire burn area, Lake Creek Coulee.

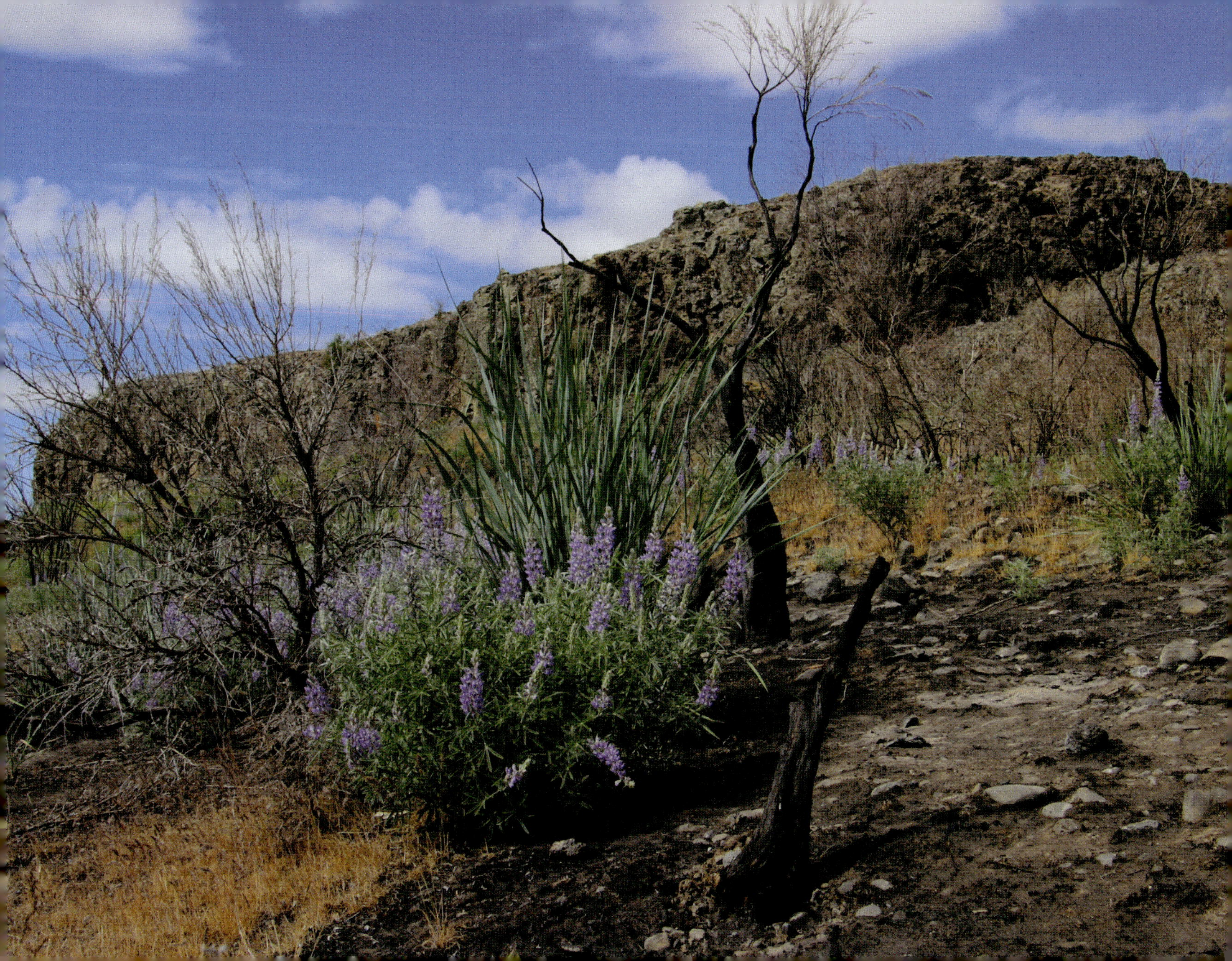

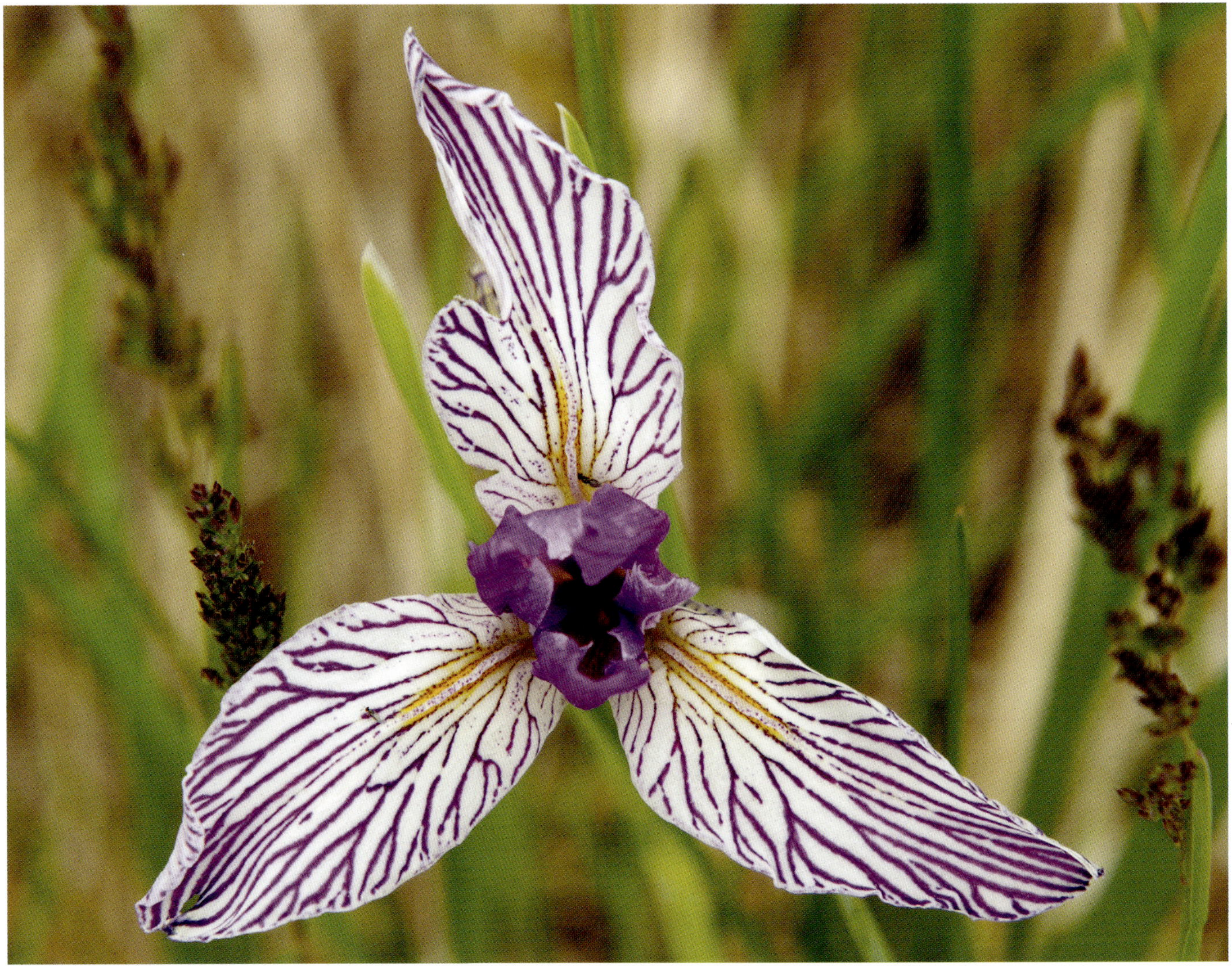

Left: Rocky Mountain iris. *Right:* Palisade and Saddle Mountains, Grant County.

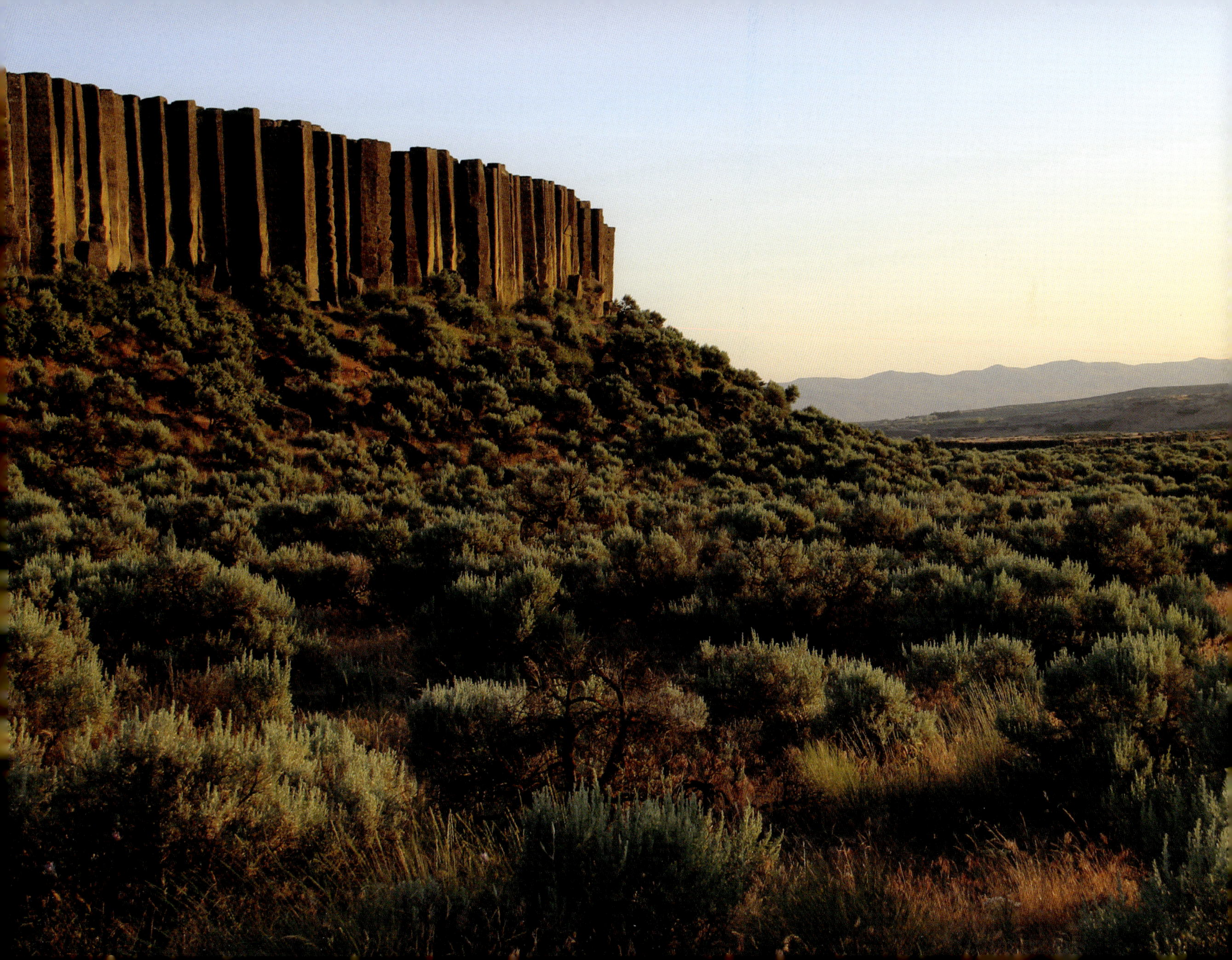

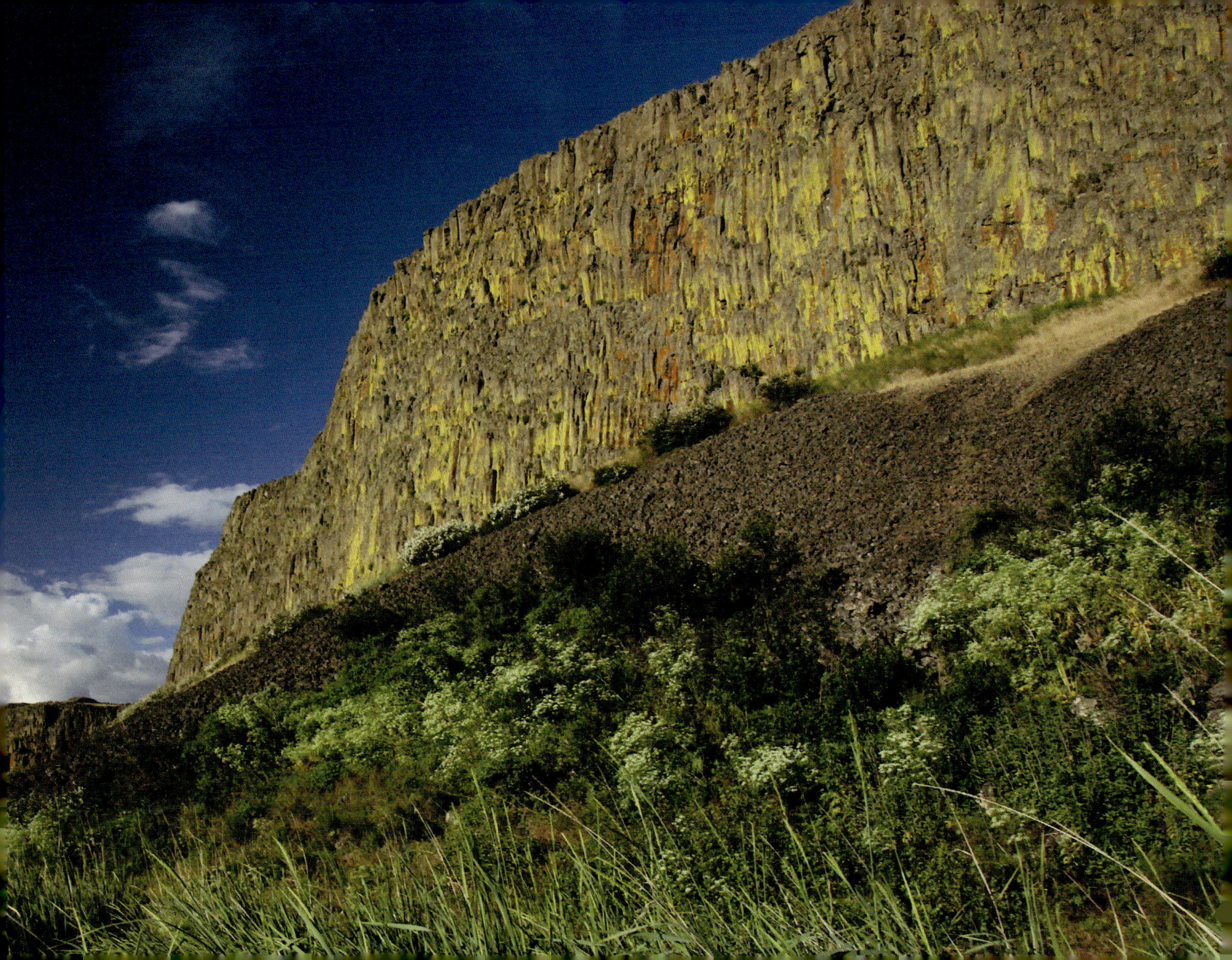

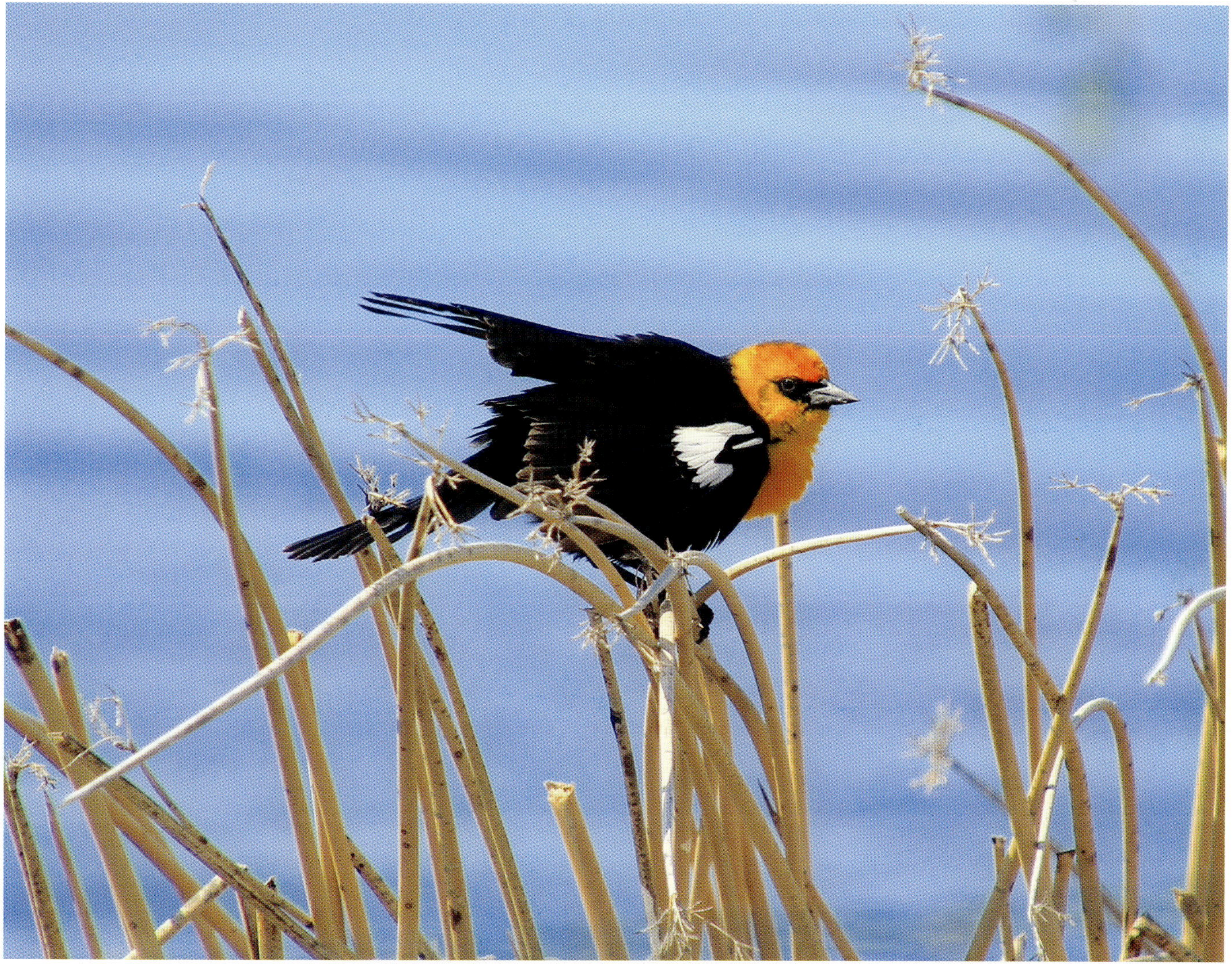

Left: Basalt butte above Rock Creek at Escure Ranch, Whitman County. *Right:* Yellow-headed blackbird in scabland marsh.

CHAPTER VII.

Someone Tell Othello

One way to leave Harrington (pop. 424) is to follow Coffee Pot Road into the vast open spaces of central Lincoln County. After a brief jog to the north, the shoulder-less pavement makes a westward bead through a patchwork of gently rolling, dryland farm fields. The asphalt doesn't bother to bend as it crosses buff and treeless terrain that gets flatter by the mile and appears endless in all directions.

This all changes, in the space of a few blinks, when the road reaches Lake Creek Coulee, a main artery of the Telford scabland tract. As with every other boundary between arable soil and scabland, it appears as though the earth is suddenly changing clothes. It's just that here the transition is especially dramatic, as Coffee Pot Road descends toward the Lake Creek wetlands. Almost immediately, a lichen-painted brow of Wanapum basalt appears and reflects morning sunlight like a colossal, natural billboard. If it happens to be a morning in spring or early summer, a cacophony of birdsong rises from a sprawling marsh at the base of the basalt wall.

Even in January, and perhaps especially so, Coffee Pot Road is a thin line to solitude. On maps and satellite photos, the Telford tract looks like a box jellyfish, with US Route 2 streaking across its forehead, its tentacles of flood-scoured pathways drifting from Bachelor Prairie in the north down toward Odessa (pop. 876) and the Crab Creek drainage to the southwest.

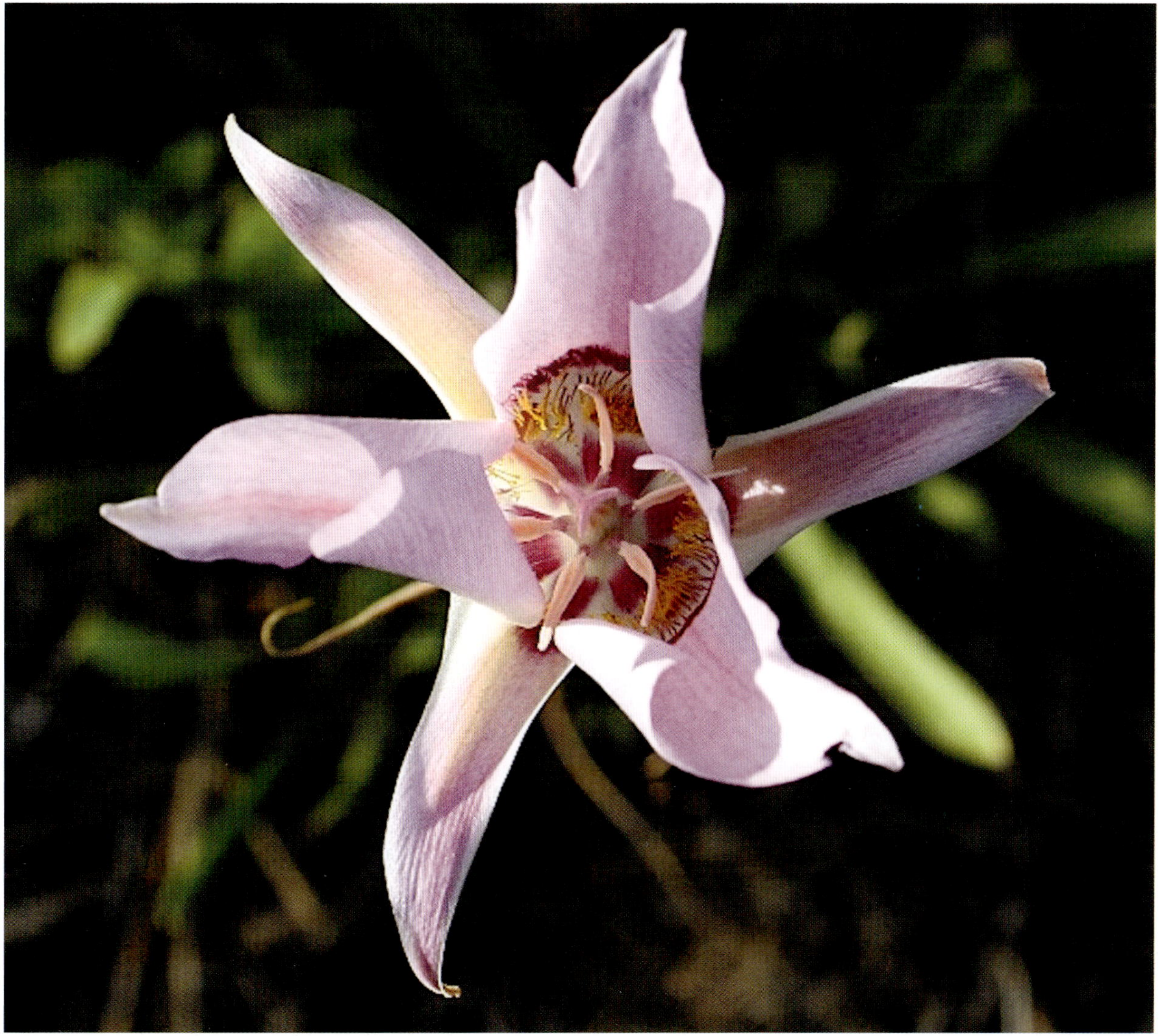

Green-banded mariposa lily.

On the clearest of days, from high points beyond Lake Creek Coulee, the snowcapped front of the Cascades is discernible on the distant lip of the horizon. Even this fleeting view may remind you just how far you are from anything else, especially knots of humans. With barely more than five people per square mile, Lincoln County is one of the more sparsely populated counties in America.

One of my favorite memories is traveling with my then-16-year-old son as we explored Lake Creek Coulee on the next to last day of December, when the sun was out but the temperature was still below freezing. We'd already encountered a large herd of mule deer when we approached Coffeepot Lake on foot. Even before we could see the lake we heard the casual honks of socializing waterfowl.

Devin climbed directly up a steep, rocky ridge between us and the lake while I walked downslope toward a gap in the phalanx of basalt. He got to the top just a few moments before I came around my corner. When he popped his head up, a swirling mass of geese took flight. The ensuing roar was the collected sound from hundreds of wings, and it so shattered the wintry silence that we were both pleasantly stunned. A minute later, all that remained on the thin blue ice of the lake was a lone swan. Devin told me later he'd never experienced anything like it.

The Telford tract is the least visited of the three main scabland pathways. It lacks cavernous, postcard venues such as Palouse Falls to the east or Dry Falls to the west, where, on warm days, crowds gather at a portable concession stand and music spills from car stereos. But whatever the Telford is missing in terms of grandeur or amenities it makes up for in wildlife and brain-teasing nuances.

Take, for example, the unusual craters that complicate the landscape near Odessa. What sets them apart is their curious multi-ring forms. The prevailing theory is the concentric walls were cast long ago as one or more massive lava flows came in contact with surface water. The resulting steam and lava explosions involved a symmetric venting of energy, pushing up circular "dikes" in the cooling basalt. When the ice age floods arrived 15 million years later, the violent rinsing removed the topsoil and the weaker segments of basalt—leaving the more resilient, ringed dike structures largely in place.

A constellation of the ringed craters is within a mile or so of where Coffee Pot Road connects to Washington Route 21, about five miles north of Odessa. The Odessa Craters Trail connects the Amphitheater, Rock Rose, and Wild Garden Craters. Especially in late spring, the craters and the rocky terrain around them are piped in wildflowers and teeming with wildlife, including a loose delegation of rattlesnakes. The cliff-like south wall of Amphitheater Crater is so densely coated with yellow and orange lichens that seeing it for the first time may induce a gasp. For those without other appointments, a lone bench is anchored near the center of the crater.

After excavating Lake Creek Coulee and unearthing the Odessa craters, the Telford floodwaters took a westward turn into the Crab Creek drainage. The great hydraulic plow then followed the creek's course as it moved downgradient toward Moses Lake and into the Quincy Basin. There, floodwaters from both the Telford and Grand Coulee tracts pooled before their gravity-induced rampage sent them toward Pasco and Wallula Gap.

The eastern outlet for the Quincy Basin is the Drumheller Channels, a dramatic maze of buttes and channels north of Othello. The northern portion of the Drumheller is now beneath the man-made Potholes Reservoir (not to be confused with Potholes Coulee, which is also in Grant County but thirty miles to the northwest). But the southern part of the channels is dry and largely intact, encompassed by the Columbia National Wildlife Refuge.

Bretz touted the Drumheller as "the most spectacular tract of butte and basin scabland" in the region, possessing "an almost unbelievable labyrinth of anastomosing channels, rock basins, and small abandoned cataracts." The non-submerged portion of the Drumheller is fifteen miles long and roughly half as wide.

A few years ago, my old friend and journalism partner, Larry Shook, went with me on a long hike through the channels. Our destination was a climbable butte skirted with a visually stunning palisade of Elephant Mountain basalt columns. I realized later that this was neither the easiest nor shortest way to approach the majestic butte. But maybe it was still the best way, hiking through the "labyrinth" that Bretz described, winding our way around, and sometimes over, bare ridges that still look freshly scoured by the great floods. Midway through our walk we heard the haunting calls of sandhill cranes as they circled in the powder-blue sky above us. A few minutes later, as we surmounted another ridge, we could see our destination—a surreal wall of thin rock columns like a big, upright stack of pencils.

From atop the butte we could look back at the terrain we'd just crossed, in the direction from which the floods arrived. From this vantage, it looks like Mars, especially in summer, when the landscape has turned from green to ocher, punctuated with frozen black waves of exposed basalt. Because there are no trees, the rawness of the landscape is all the more visceral, and it's easy to imagine another devastating flood arriving within the hour. A whimsical tagline for one of the photos I took from the butte reads: "Someone Tell Othello."

The terrain is as Bretz reported a century ago, but I was also struck by what we hadn't seen. There were no other people in sight, and the only man-made debris I'd noticed, across several miles, was a single, weathered bottle cap.

I've been back to the Drumheller several times since, in all seasons, most often alone but sometimes with companions who'd never been there before. It isn't until the moisture of spring gets sucked away by warmth and winds that my favorite desert wildflower, the green-banded mariposa lily, comes into bloom. It is a showy, mystical plant that only appears east of the Cascades and often in places that, by late May, look panfried and barren. I was scouting another butte in the Drumheller one early summer evening, waiting for the light to ripen, and had the pleasure to notice that I was surrounded by dozens of the blooming lilies, more than I'd ever seen in one place before. The life-giving ribbon of Crab Creek was flowing through tall reeds in the valley below me while, above, a languid great egret was flying in the same direction, roughly toward Mabton.

Right: Spring storm, central Lincoln County.

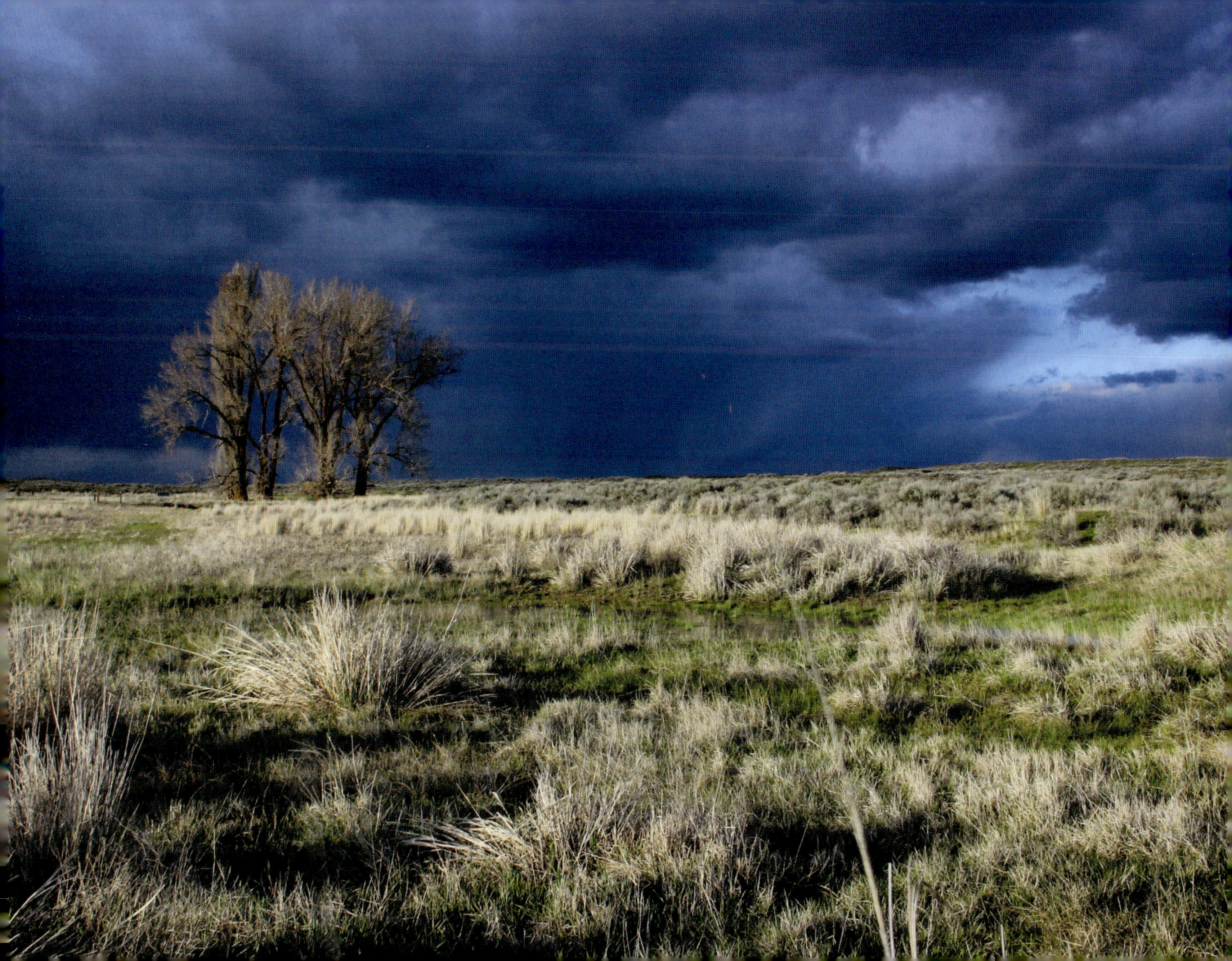

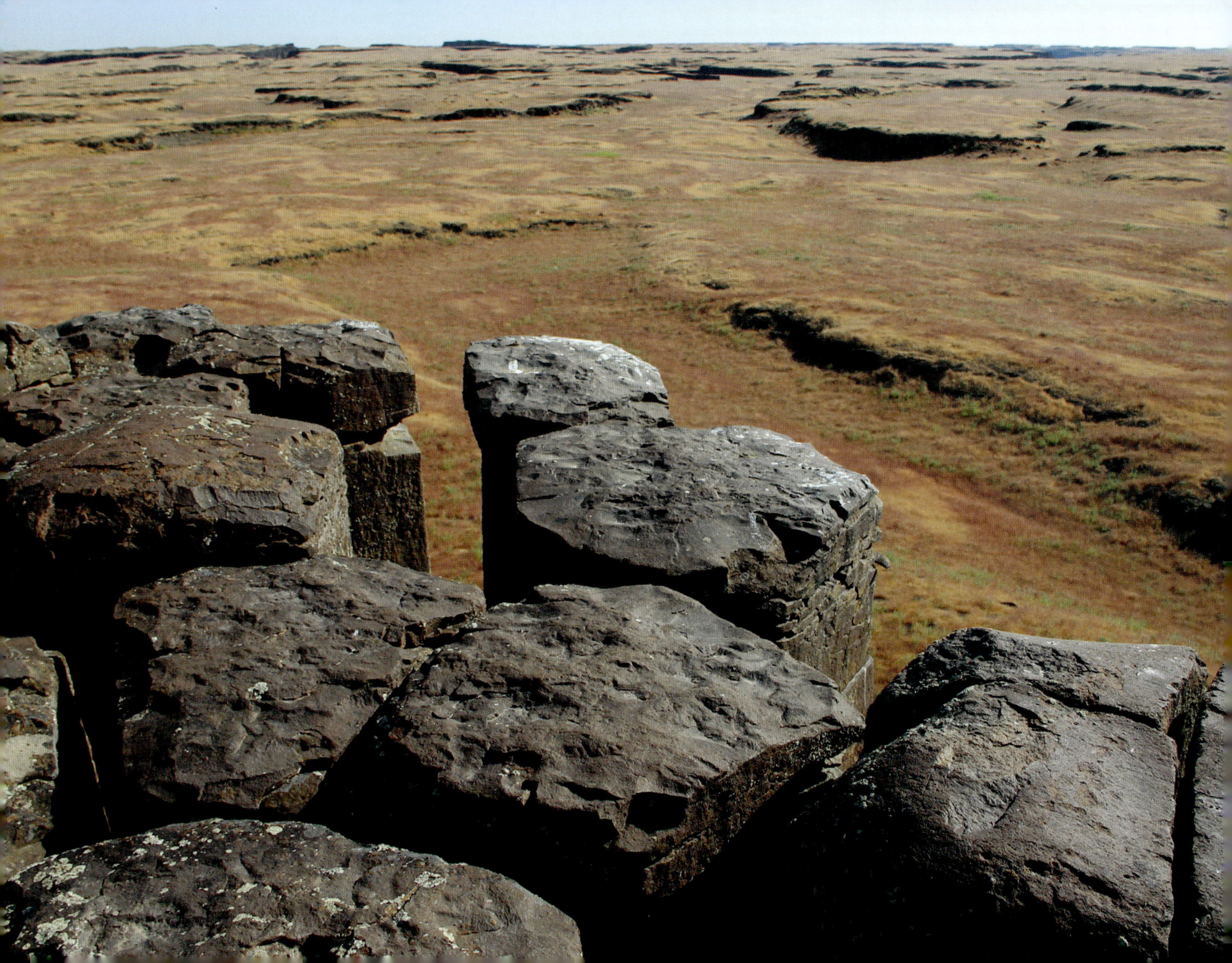

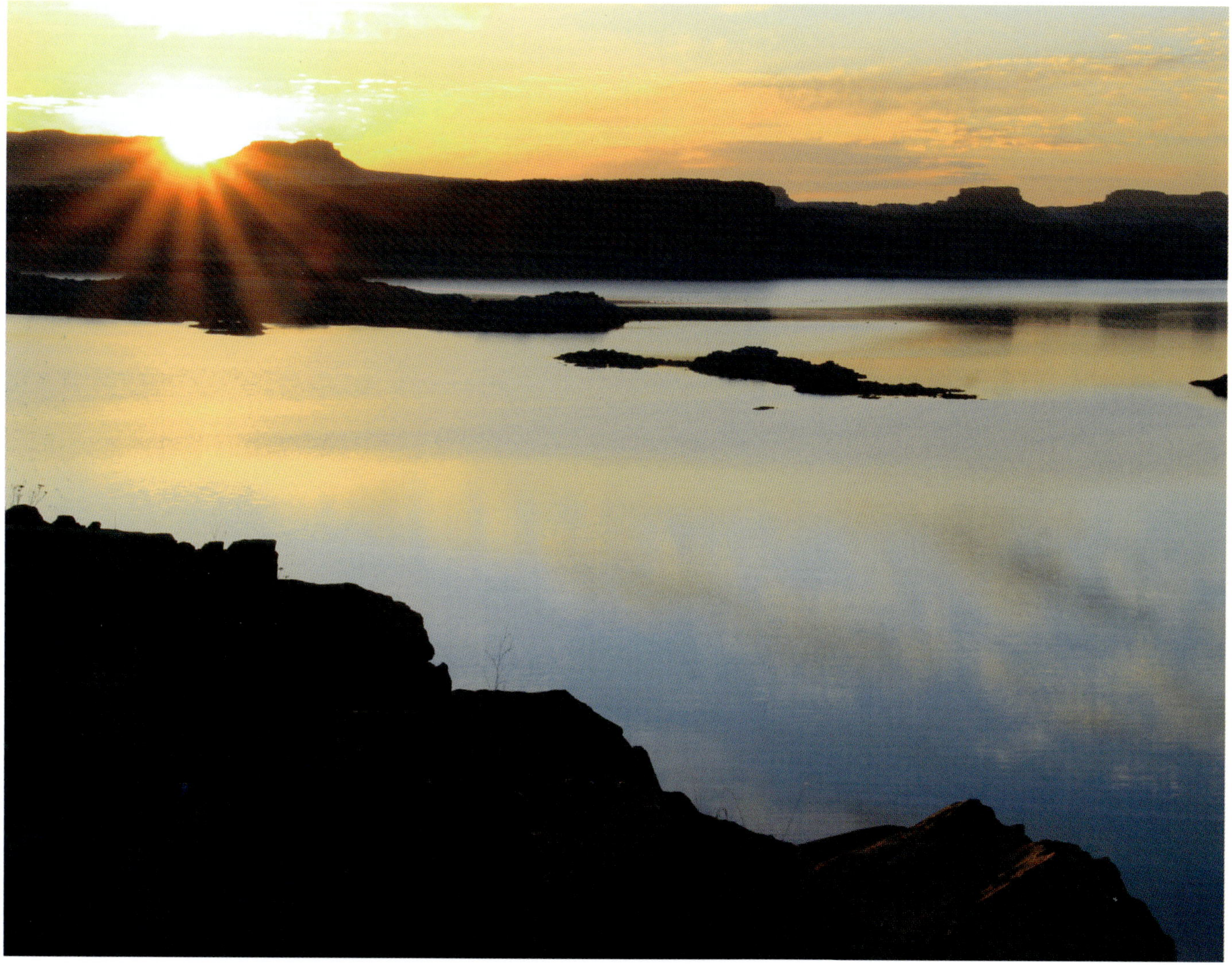

Left: Elephant Mountain palisade, Drumheller Channels. *Right:* Sunrise at Soda Lake, Drumheller Channels.

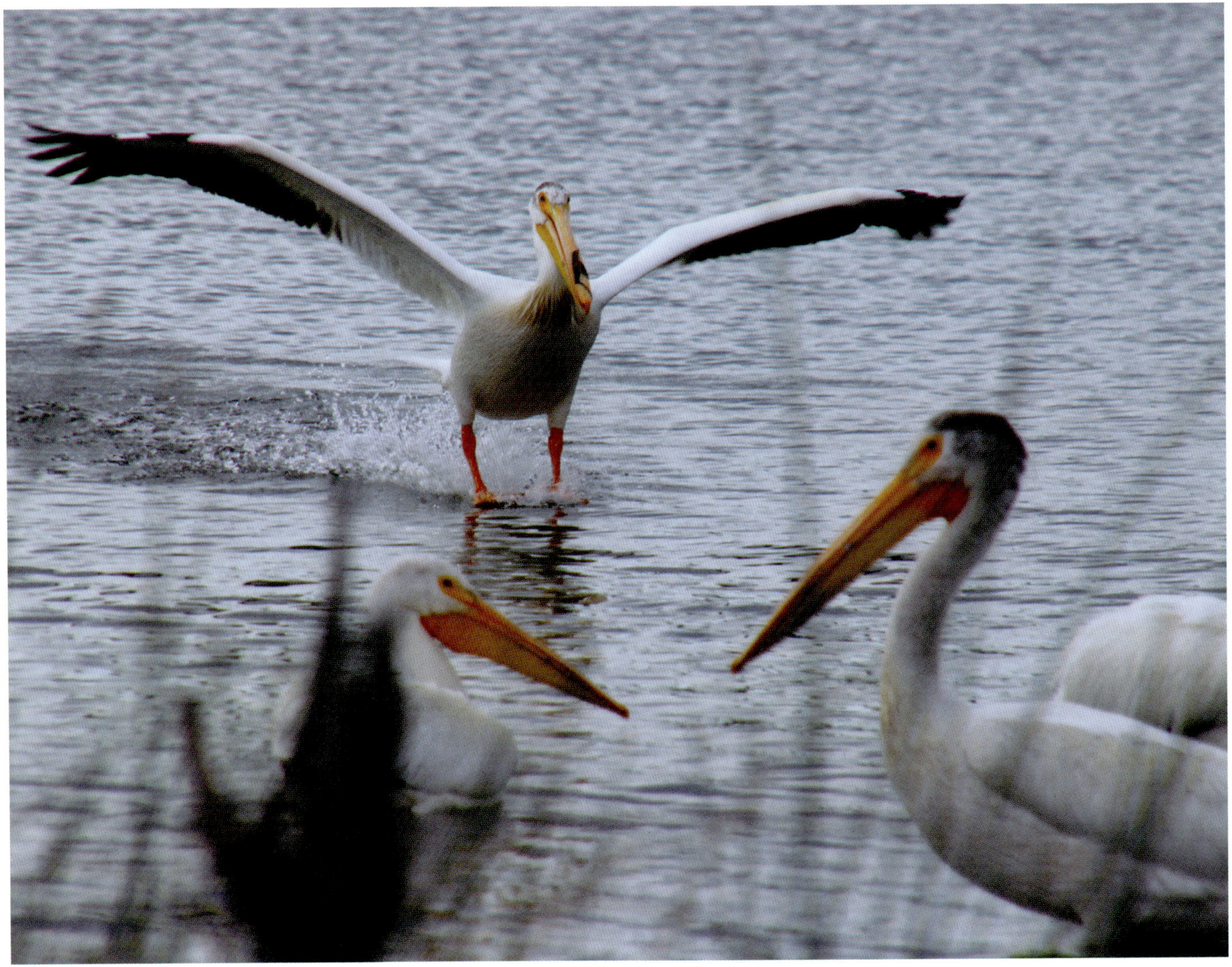

Left: American white pelicans at Clear Lake, Spokane County. *Right:* Lichen-coated wall near Odessa.

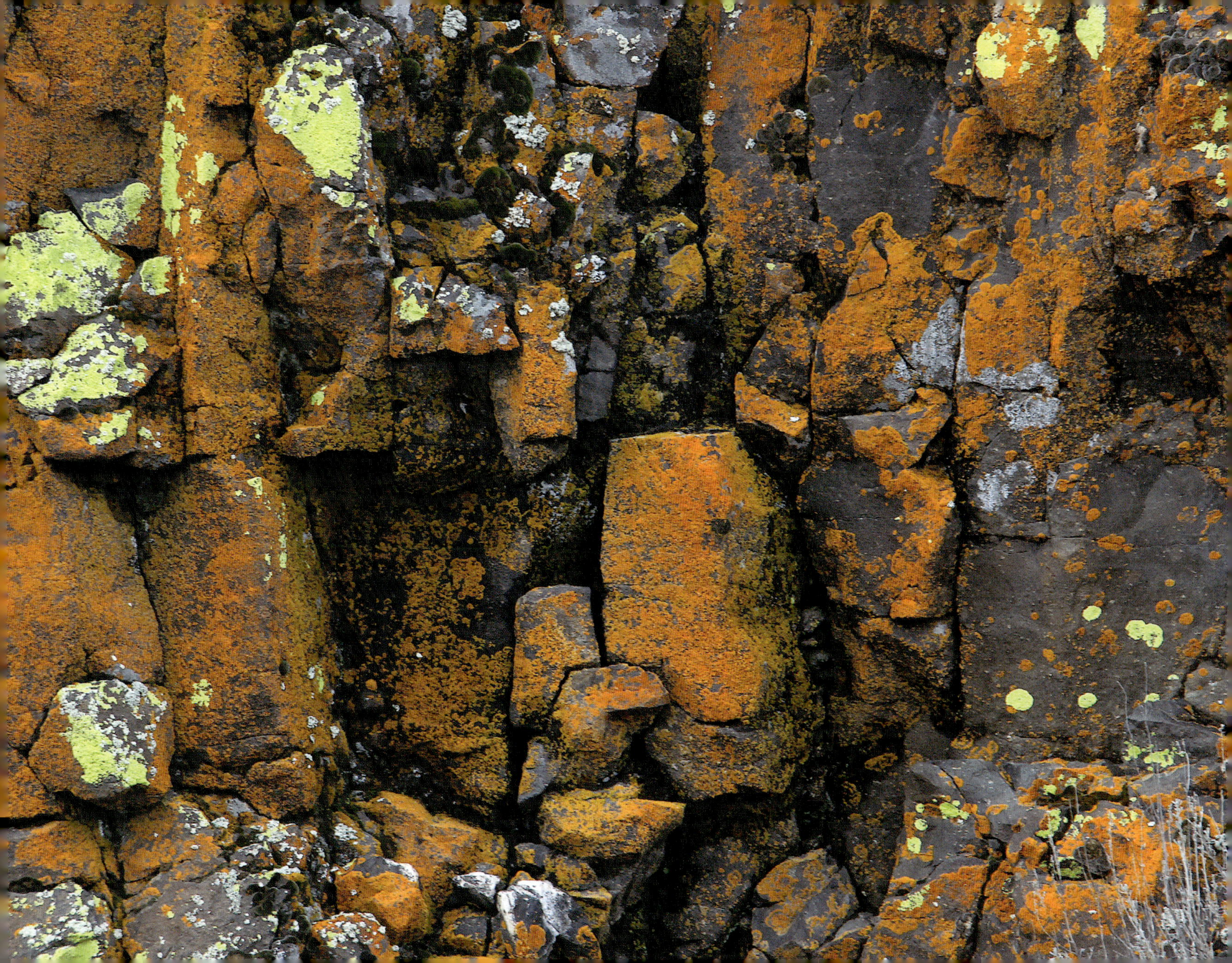

CHAPTER VIII.

Grand Chasms

The first time I sought refuge in Northrup Canyon was on a late summer day as I was on my way home from a funeral. I was grieving the loss of one of my mentors and resisting the pull of familiar surroundings and distractions, which is where the highway would have taken me.

The fork in the road was just beyond Coulee City, where US Route 2 crosses the long, earthen dam separating Banks Lake from Dry Falls and the corrugated grandeur of lower Grand Coulee. Instead of continuing eastward toward Spokane, I followed Washington Route 155 northward along the lake and deep into the throat of upper Grand Coulee.

Unlike the lower coulee—which braids into a labyrinth of interconnected canyons several miles across—the upper coulee leaves Coulee City as a U-shaped half-pipe. The coulee is natural, but the twenty-seven-mile-long reservoir that floods the trough is nothing of the sort. Banks Lake is Columbia River water that has been pumped uphill—nearly 300 feet—for irrigation storage. On a calm day, the surface of the two-mile-wide reservoir becomes a reflecting pool for the cliffs that tower above it. To correctly imagine the height of the cliffs, it may help to envision the Space Needle holding up the Statue of Liberty, including the pedestal.

Midway between Coulee City and Grand Coulee Dam, the highway ramps up a talus slope from where it offers a panoramic view toward Electric City (pop. 1,001). The island blade of Steamboat Rock is visible on the northern horizon.

The scale of upper Grand Coulee is so expansive you can easily lose perspective. When you're in the trough of the coulee, Steamboat Rock appears to have risen from below like an uplifted mountain. At the crest on Route 155 you can see that the signature anvil is actually just a surviving chunk of the former plain that the floodwaters tore away.

The nakedness of the rock walls and treeless landscape can also lull you into a sense that everything is exposed and in sight, that all dimensions are visible in shades of earth, water, and sky. But that's also an illusion, and Northrup Canyon is proof of that.

The memories of my old friend Bill Mitchell were like a loose pile of photographs, and leafing through them brought tears to my eyes. It was through Bill that I'd met my wife, Connie, a therapist and peace activist from South Carolina. And six years later, just hours after our first child was born, I took a picture of him laughing with delight as he held tiny Audrey in his arms. Most importantly, his life was devoted to bringing people together and helping them protect their neighborhoods from perfidy and pollution. He'd fought the good fight. At his crowded memorial on Vashon Island, people from all over the country had come to share their stories and gratitude.

I wanted to be alone for a while longer to think more deeply about what his life meant to me and my family. And that's how I wound up in Northrup Canyon, which is invisible from Route 155 but reachable on a gravel road that climbs over a ridge to the base of the canyon.

Of course, Bretz had been there decades before I hiked in. He was meticulous about using measurements to soberly insist that such epic landscapes could only be explained by catastrophic floods that overwhelmed existing drainages and created new ones on a grand scale. And still he would often address his detractors and skeptical peers indirectly, all but pleading with them to open their eyes.

"Again and again the writer has asked others and himself 'Where is the dangerously weak point in the flood hypothesis?' " he wrote in 1932. "One can stand on the brink of Northrup Canyon and aver with confidence that it can never be satisfactorily explained except as an extinct Niagara."

The questions I was wrestling with that summer afternoon had more to do with mortality, human aspiration, and transcendence than with geology. But that may be unfair to

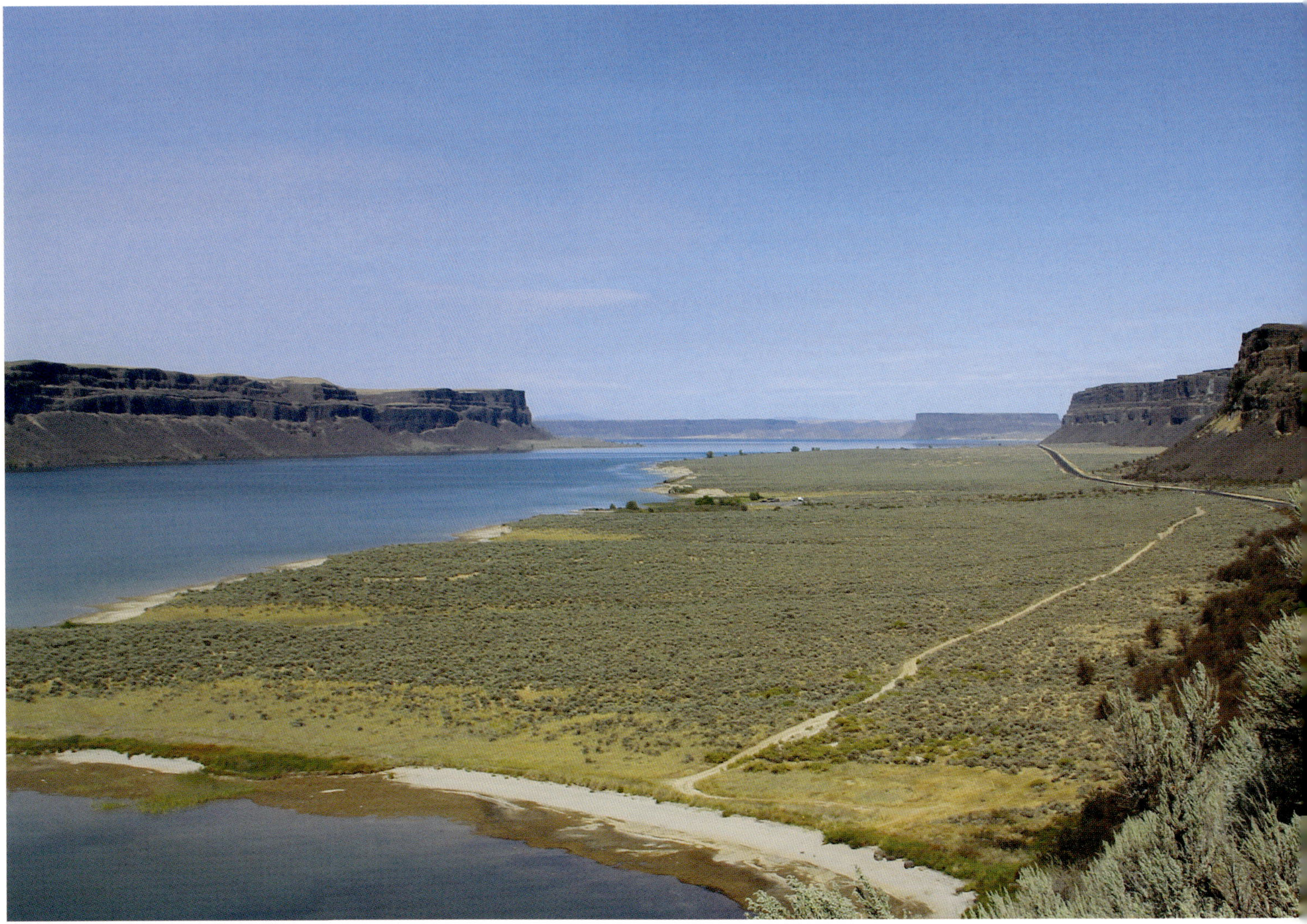

Upper Grand Coulee.

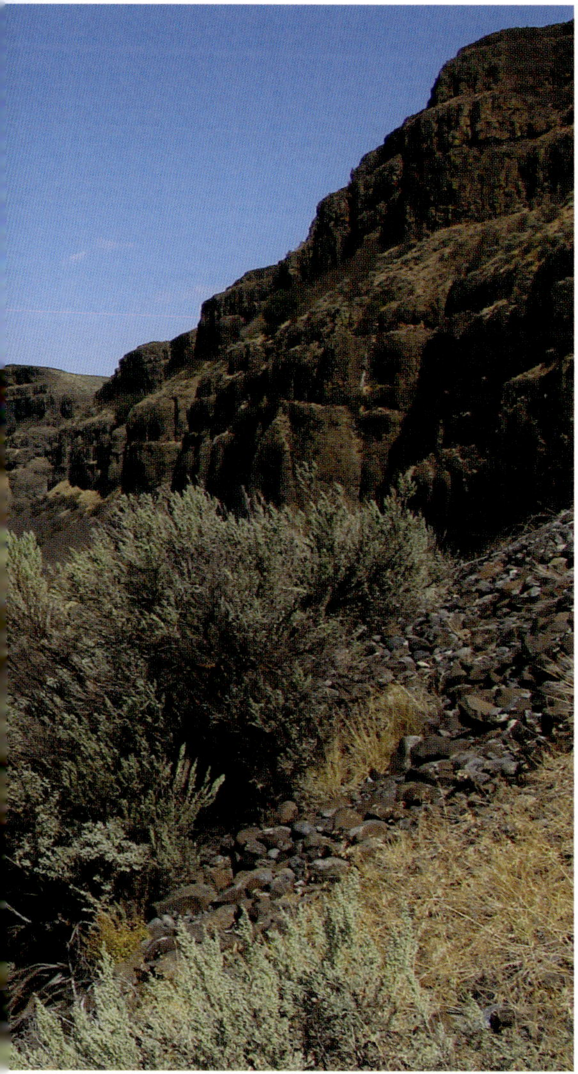

Bretz because his devotion to reason is what drove him. His legacy isn't so much about the stories in the rocks as it is about how we extract truth from a world that is as reluctant to disclose it as humans can be to accept it.

My departed friend had motivated and challenged me and given me the opportunity to prove myself. Had I rewarded his confidence and trust? Had I made the best of it? There were other questions, all of a sort that can bring you to your knees in church. My first impression upon walking into Northrup Canyon was that I had walked into a church—a magnificent natural cathedral.

From a narrow entry at a rocky bench above Banks Lake, the main trail into the well of the canyon winds through sage and pine before reaching an aspen grove wrapped in red-twig dogwood. By the time you've reached the aspens, the stunning detail in the north wall has come well into view. What's striking is the water-sculpted relief in the basalt front just beneath the canyon's battered rim. It is a long line of gnome-shaped slabs, some the size of six-story buildings, some leaning into each other, some standing alone; altogether looking like a procession of gigantic, hooded druids walking above a steep, talus slope.

The sheer power of the bedrock-gouging floods that carved the rock into what Bretz termed "an extinct Niagara" also left a mark below. The floods tore so deeply through the basalt, they exposed a gleaming, underlying formation of granitic rock—the bedrock beneath the bedrock, so to speak. Near the mouth of the canyon, a massive molar of granite—nicknamed Gibraltar Rock—towers above a picturesque equestrian area. Instead of a flat trough in the well of the canyon, what you find is a long ridge of granite in repose between the basalt walls on either side.

Because Northrup Canyon has the shape of the bone in a T-bone steak, what you see when you look down the canyon toward the east is a west-facing wall. The sense of confinement gives an impression that the canyon is a world unto itself, which, at least in a poetic sense, it is.

More broadly, the canyon's creation—merely 15,000 years ago—is a spectacular sideshow to the operatic earthworks of Grand Coulee. It is a story that unfolded over thousands of years as the Okanogan Lobe of the Cordilleran Ice Sheet advanced and retreated across the Waterville Plateau on the coulee's western flanks. The massive ice lobe was like a rail yard switch, opening and closing pathways for the inland tsunamis of floodwaters arriving from the east. The crest of Steamboat Rock is nearly 800 feet above Banks Lake, which itself is 300 feet above today's Lake Roosevelt. What that means is that the floodwaters climbed out of one gorge to create another hanging above it. At their height, the glacial floodwaters overflowed Steamboat Rock by more than a hundred feet. Just think about that.

The Bretz floods not only excavated both Grand Coulee and Northrup Canyon, but the largest pulses so inundated the Columbia River valley that they overwhelmed and reversed the flow of a major tributary, the Wenatchee River, pushing water upgradient and dropping ice-rafted boulders from Idaho as far west as Leavenworth in the Cascade Mountains—nearly twenty miles upstream.

What remains on today's landscape is a web of life-giving wetlands and dry pathways connecting natural monuments, such as Steamboat and Gibraltar. In earth time, 15,000 years is barely any time at all: less than one three-thousandth of 1 percent of the age of the planet. You can still cut yourself on the sharpness of the talus from the basalt faces scoured by the Bretz floods. And that's how it looks too—like a fresh calamity, here to remind us about the power of nature and how transient and fragile we are by comparison. Deep in these canyons, we are invited to consider how precious it is to walk into the world beyond the reach of our schedules, invoices, and smartphones—to possess our fleeting senses and the consciousness to make meaning of it all. Or at least to try.

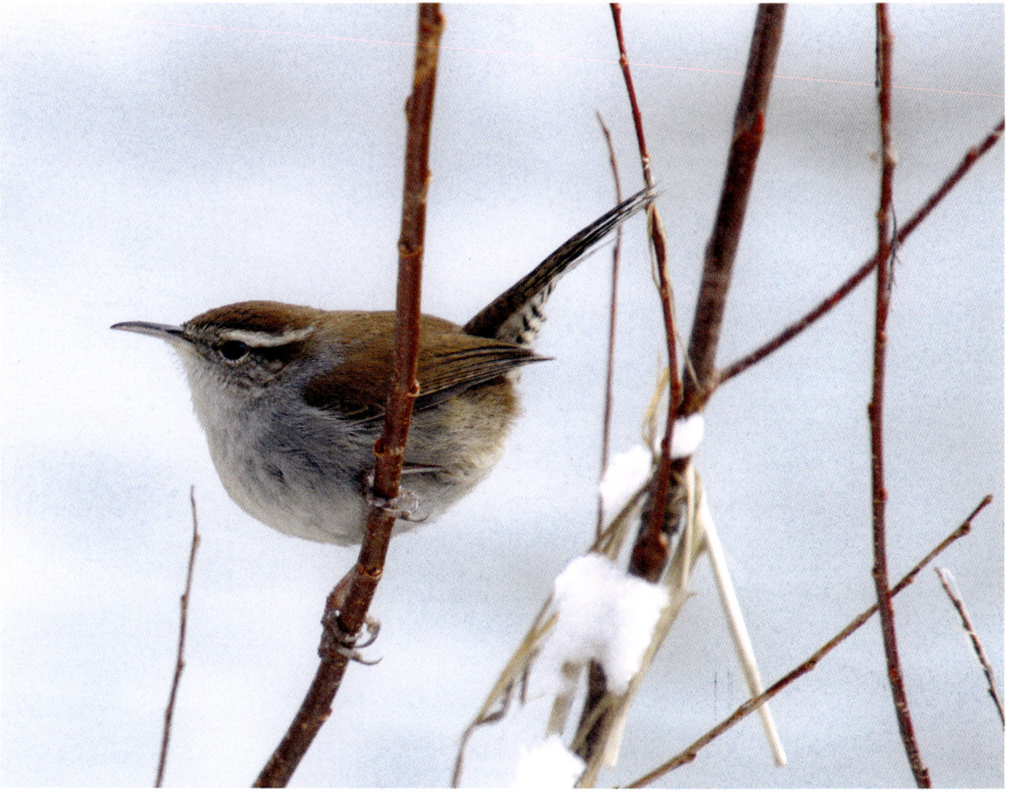

Bewick's wren in winter brush.

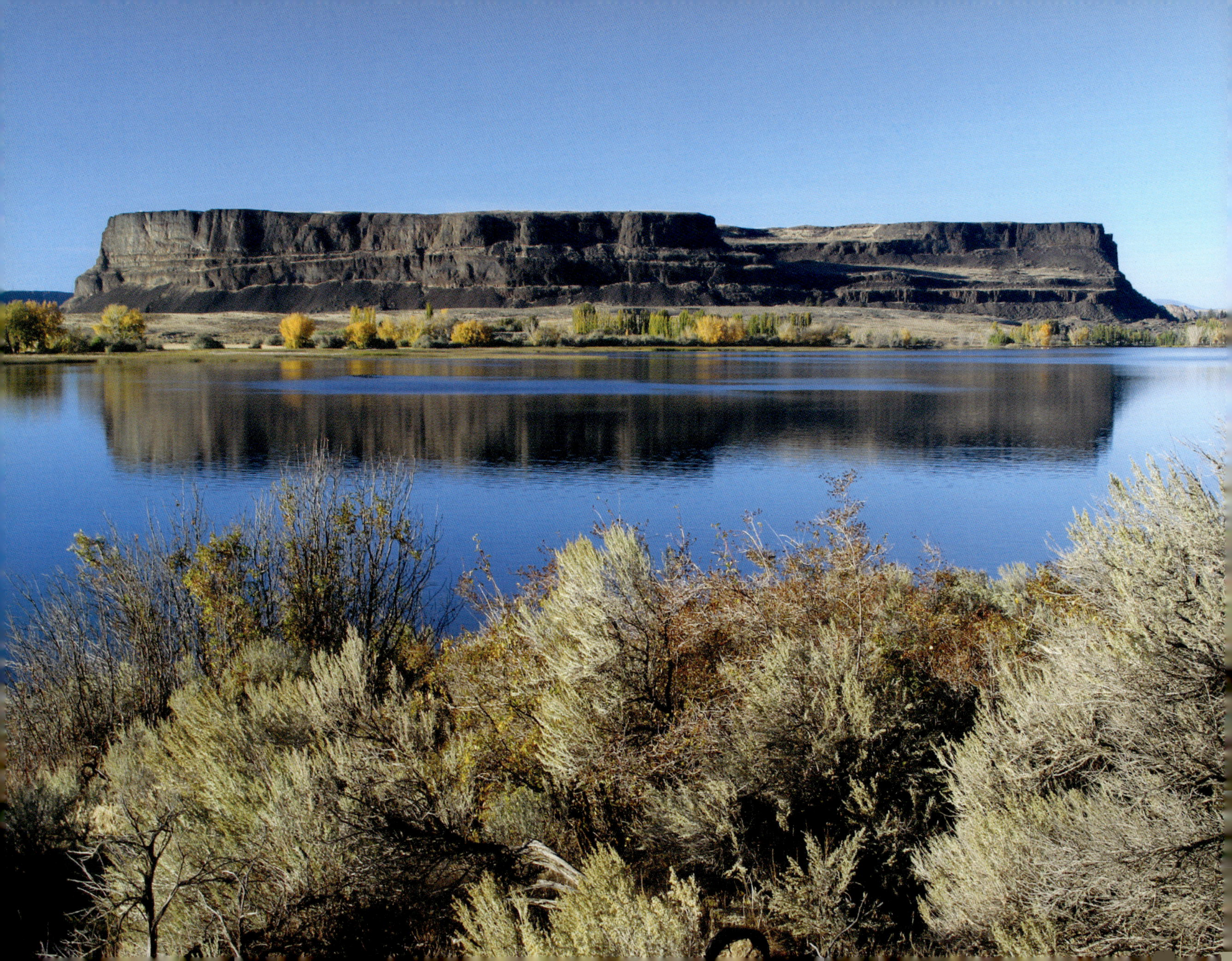

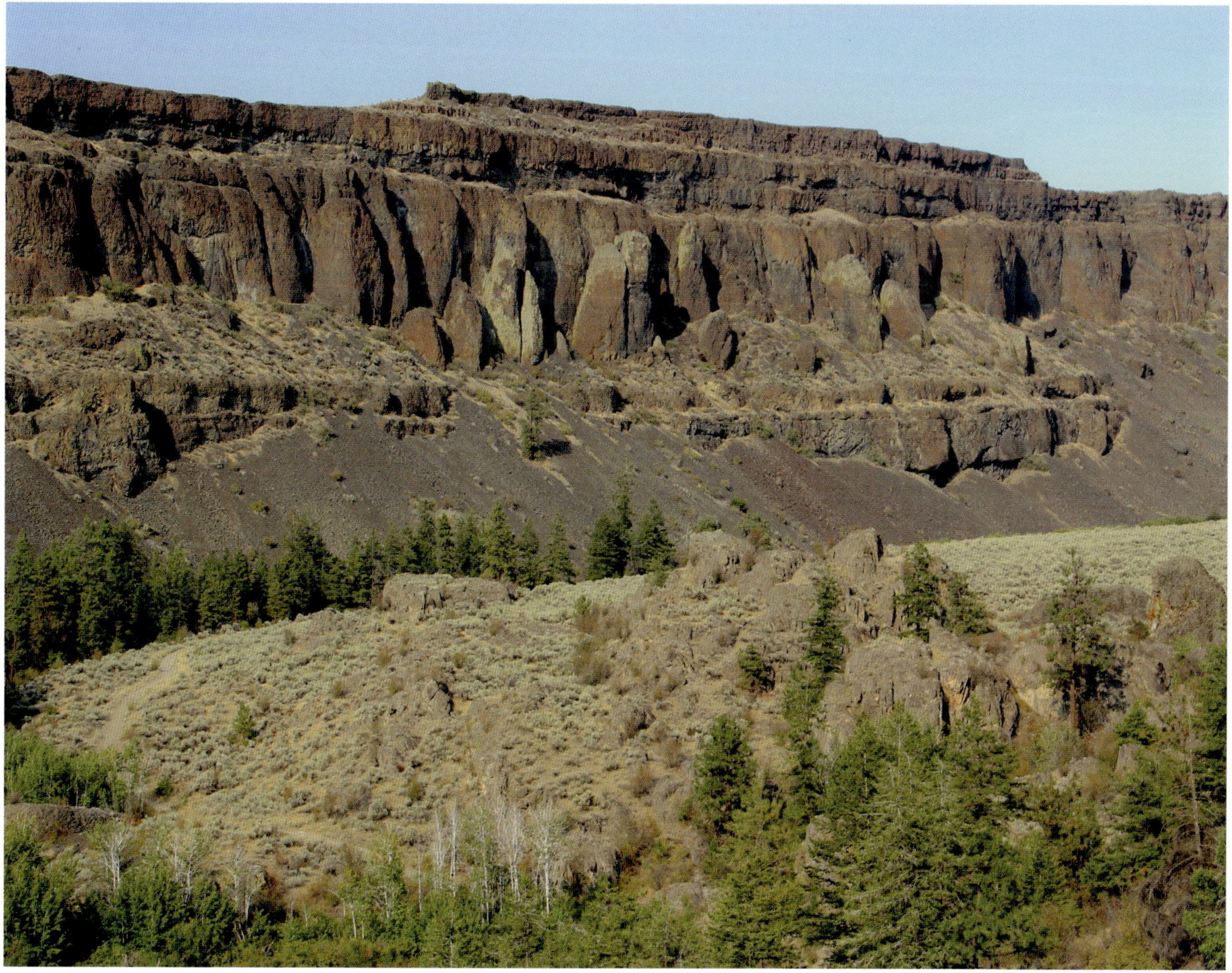

Left: Steamboat rock, Upper Grand Coulee. *Right:* North wall, Northrup Canyon.

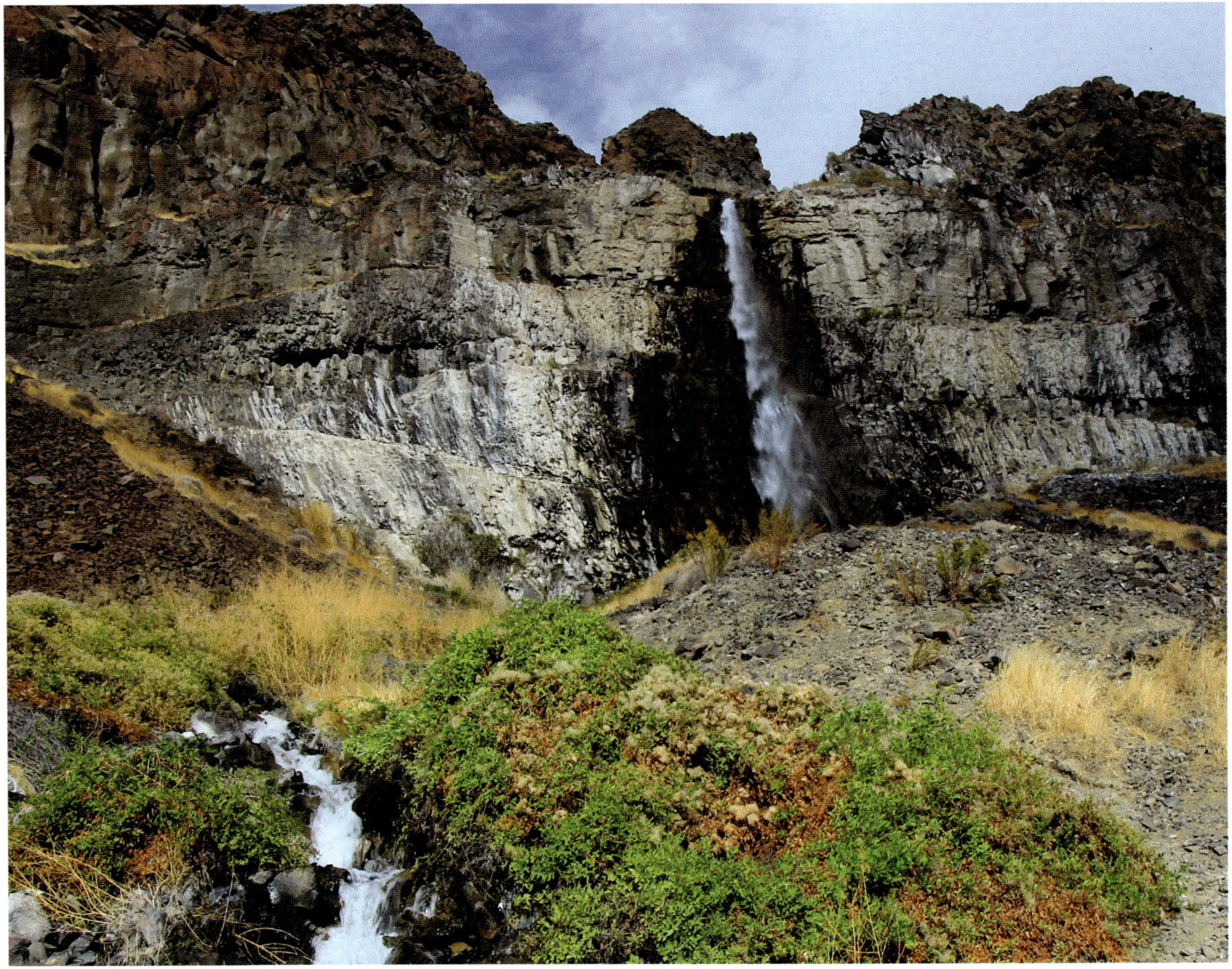

Left: Falls and stream, Frenchman Coulee. *Right:* Banks Lake, Upper Grand Coulee.

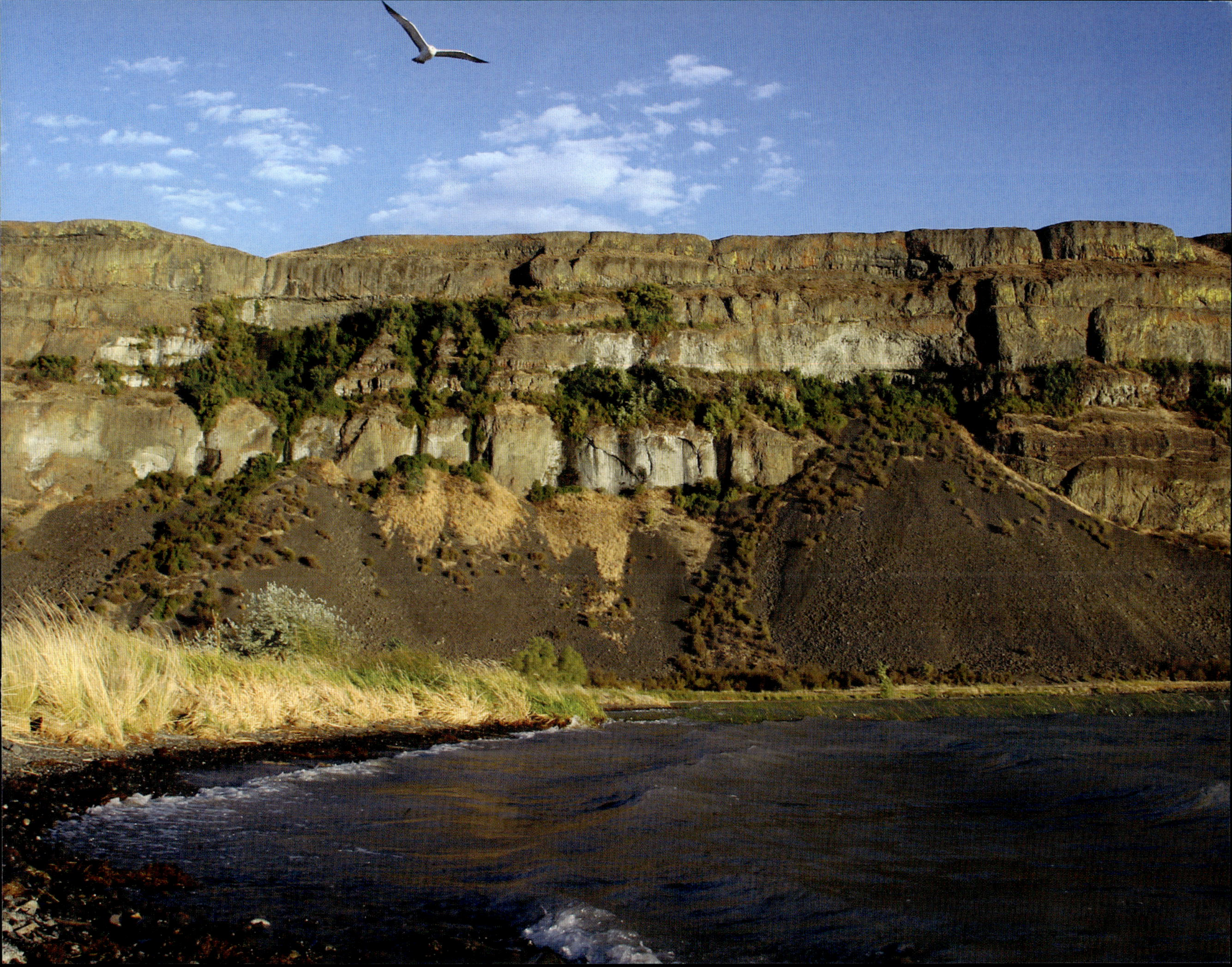

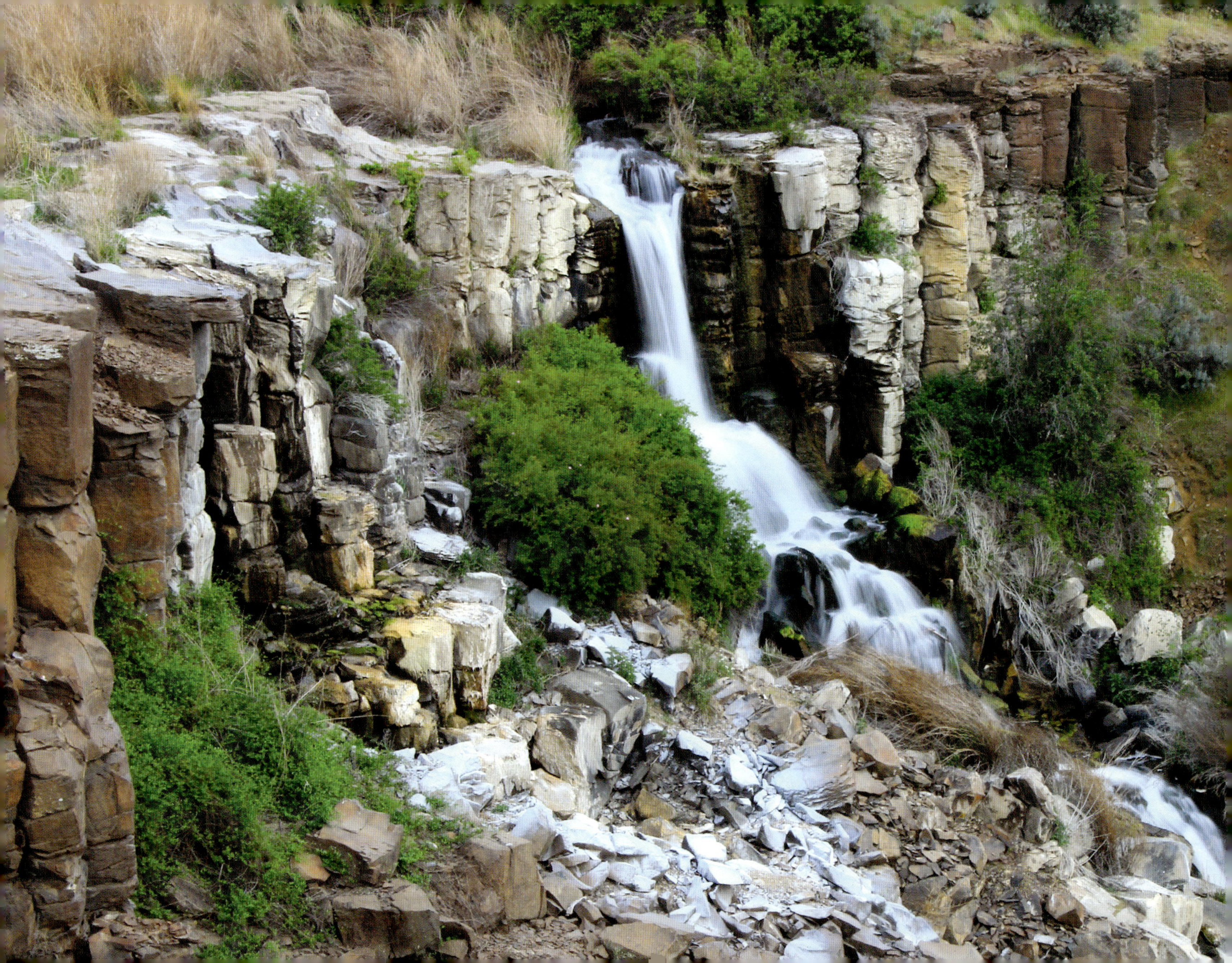

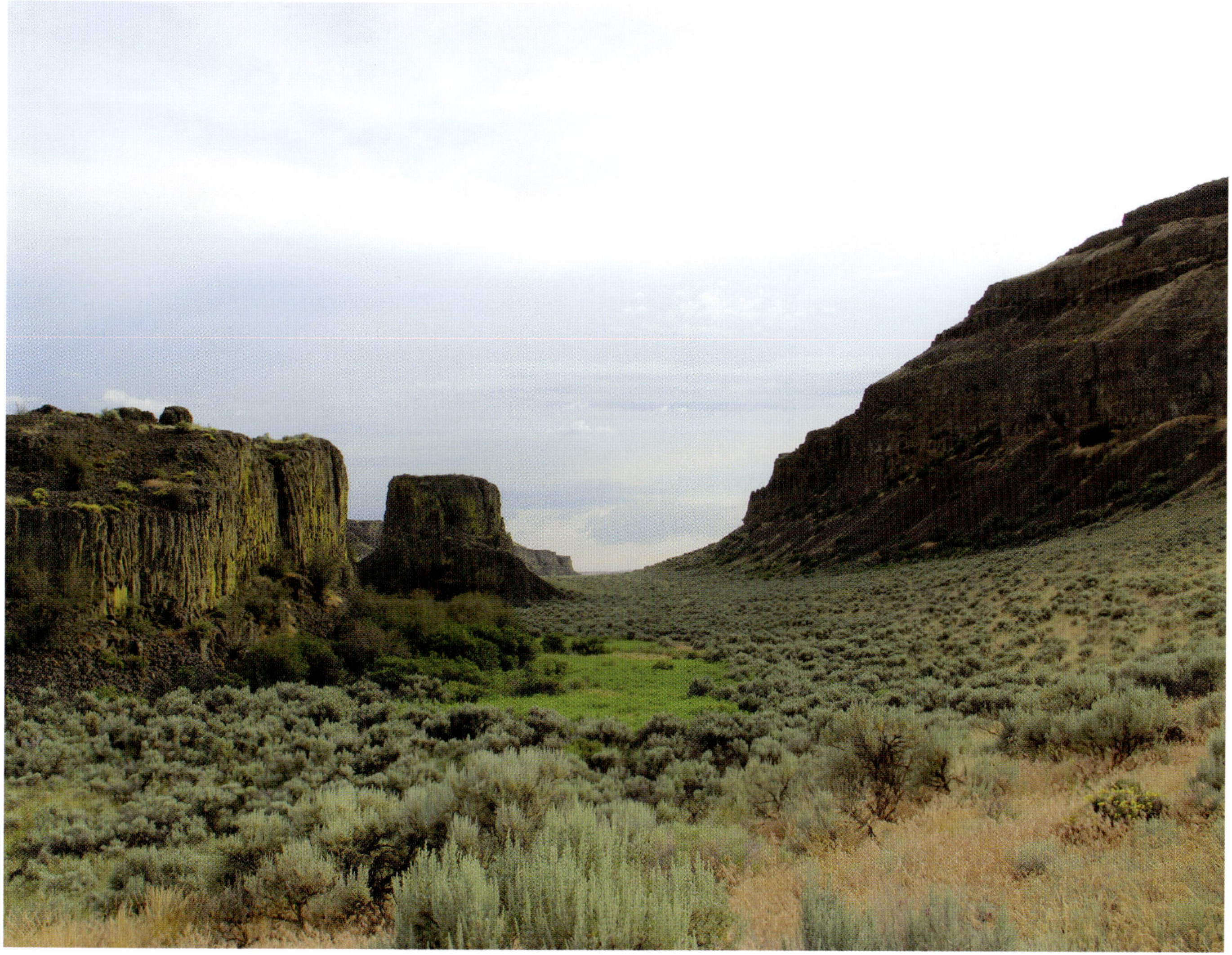

Left: Falls below Judith Pool, Potholes Coulee. *Right:* Dry Coulee, south of Coulee City.

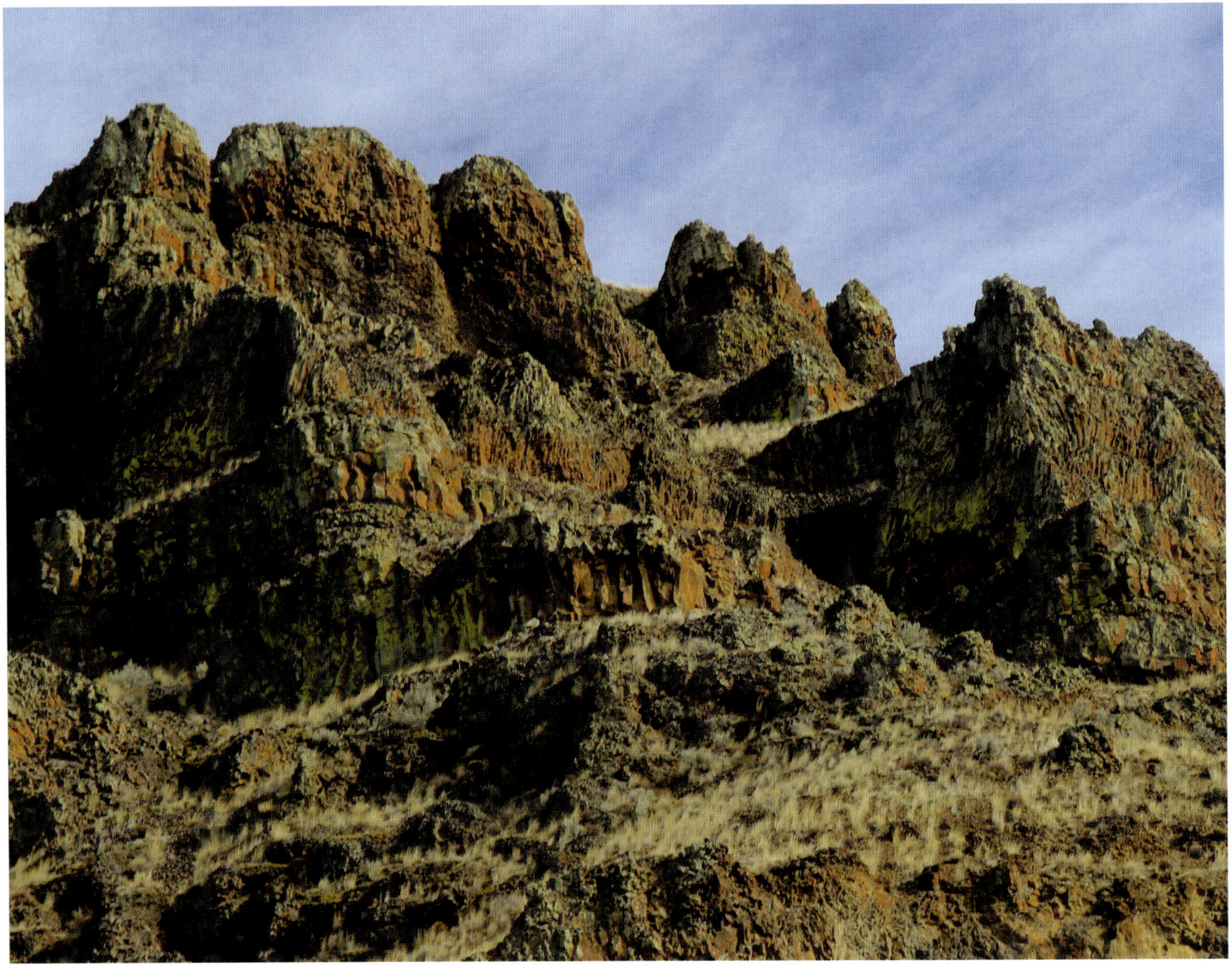

Left: Ramparts at Moses Coulee. *Right:* Palisades above Crab Creek.

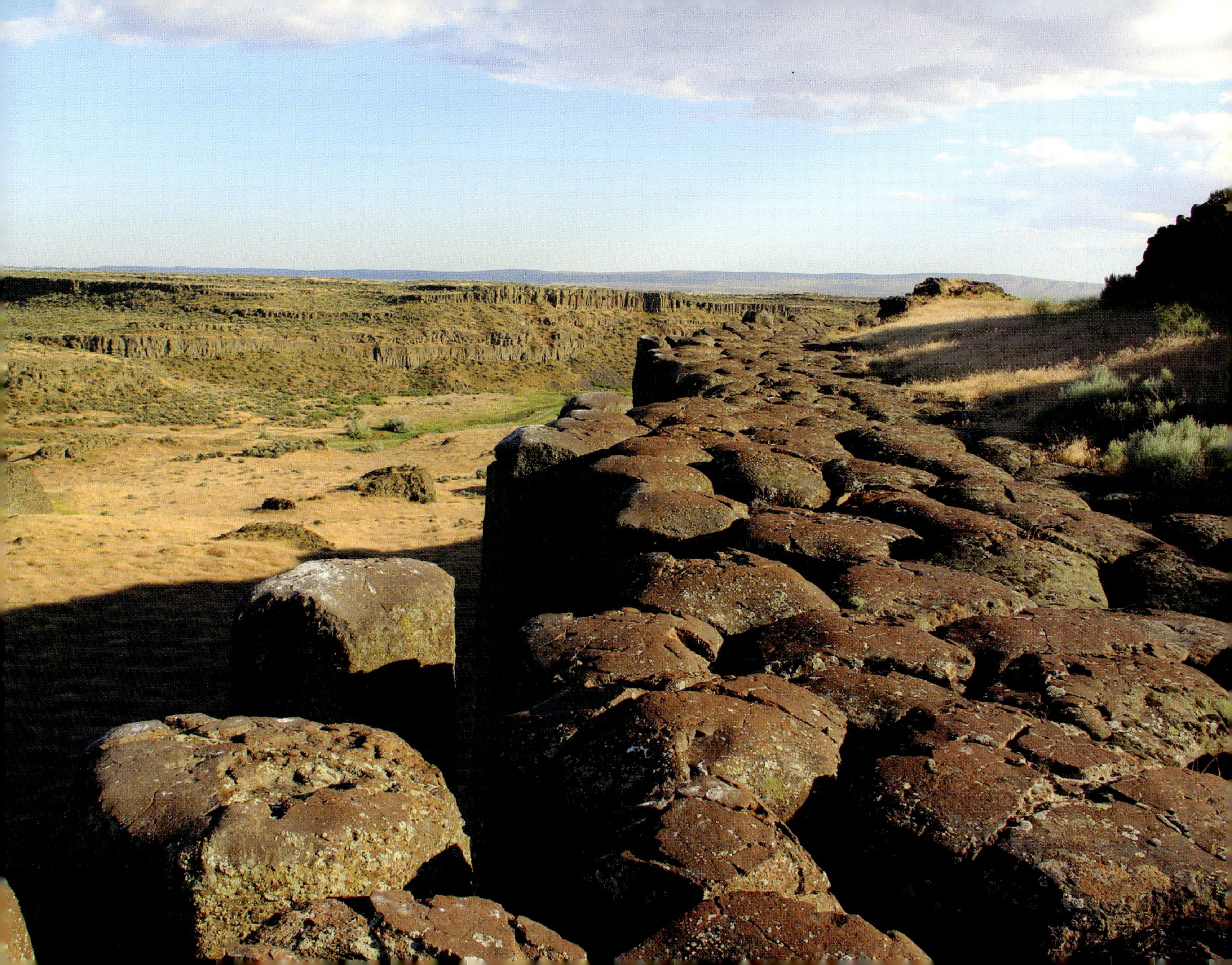

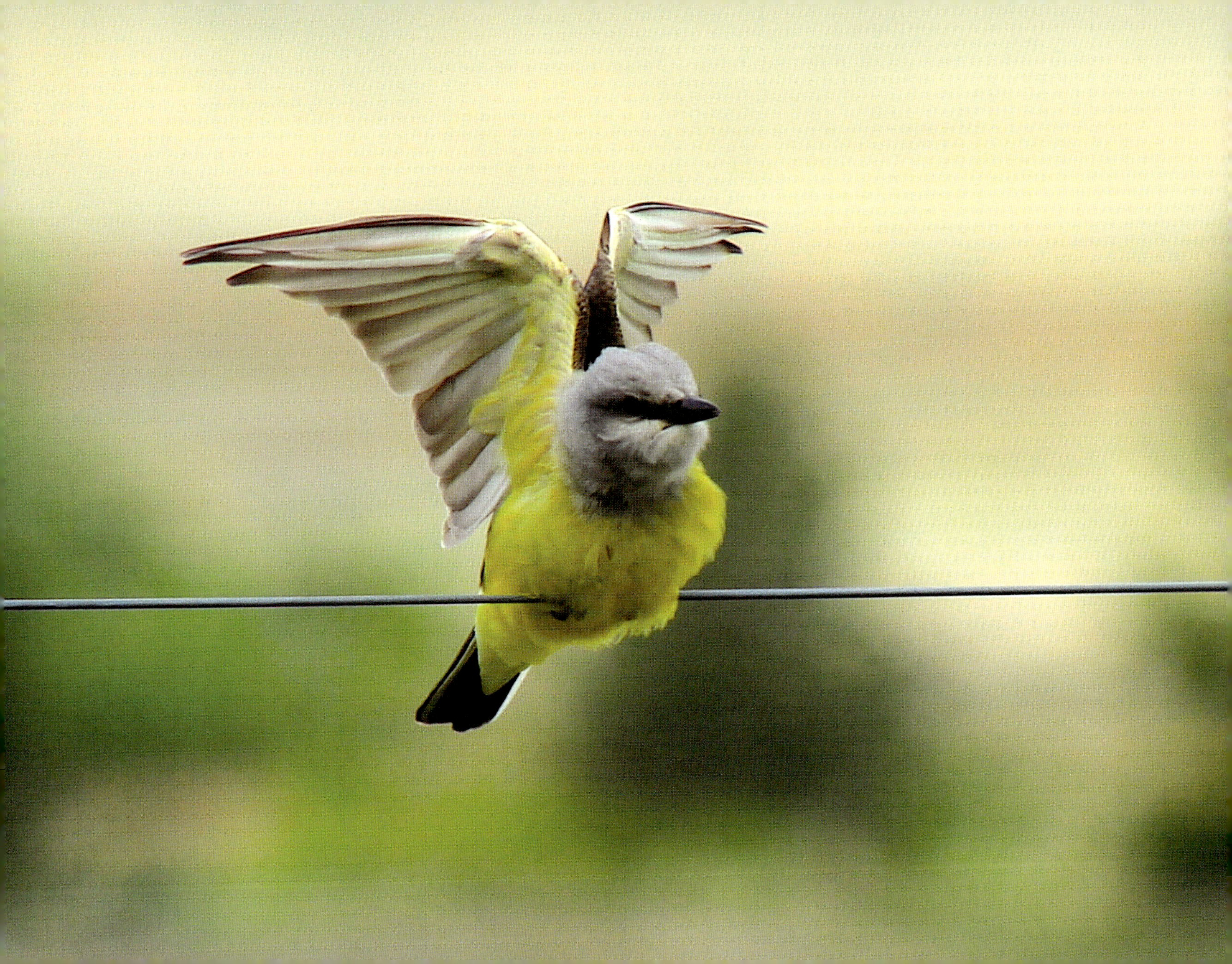

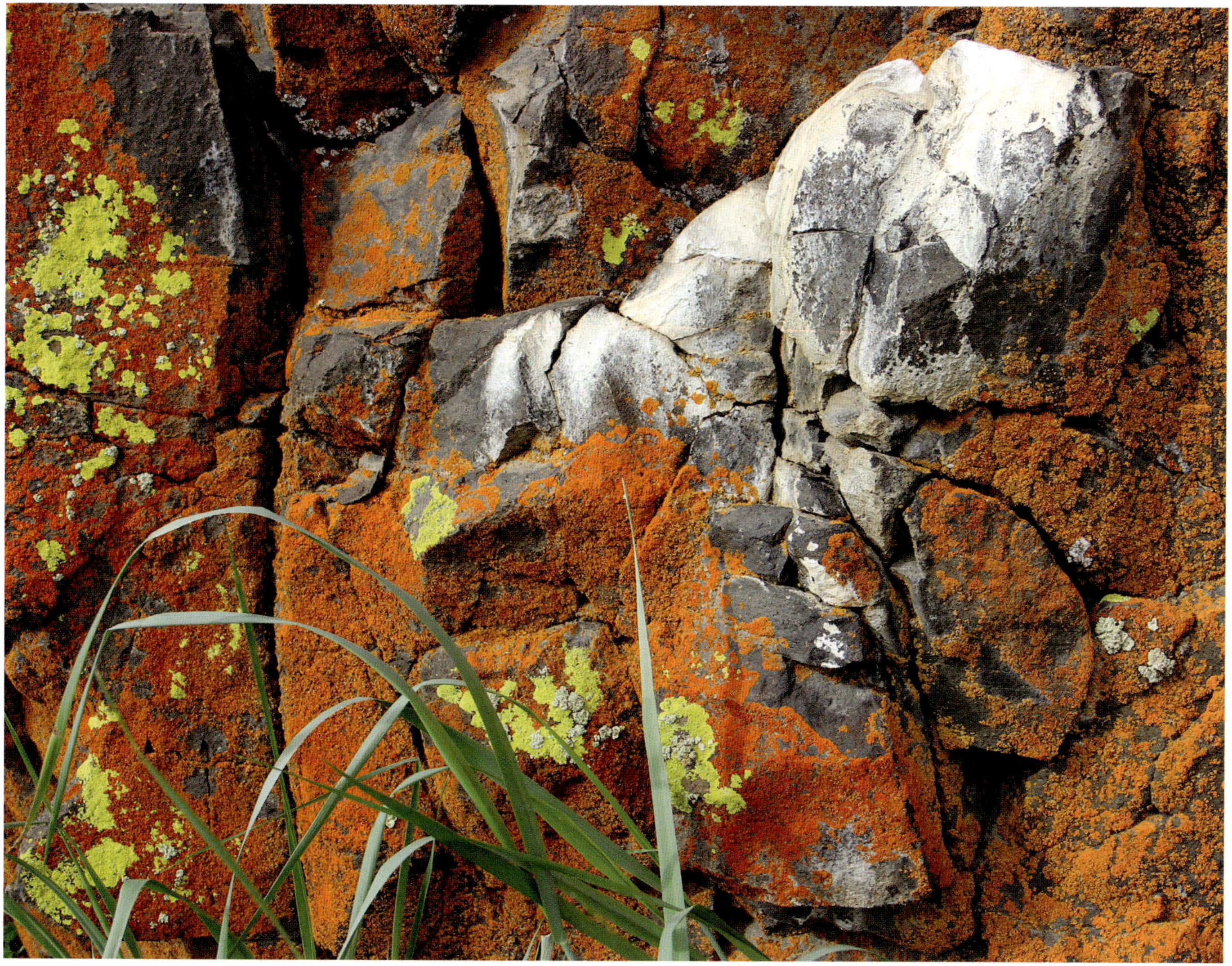

Left: Flexing Western Kingbird. *Right:* Lichens on basalt, Lincoln County.

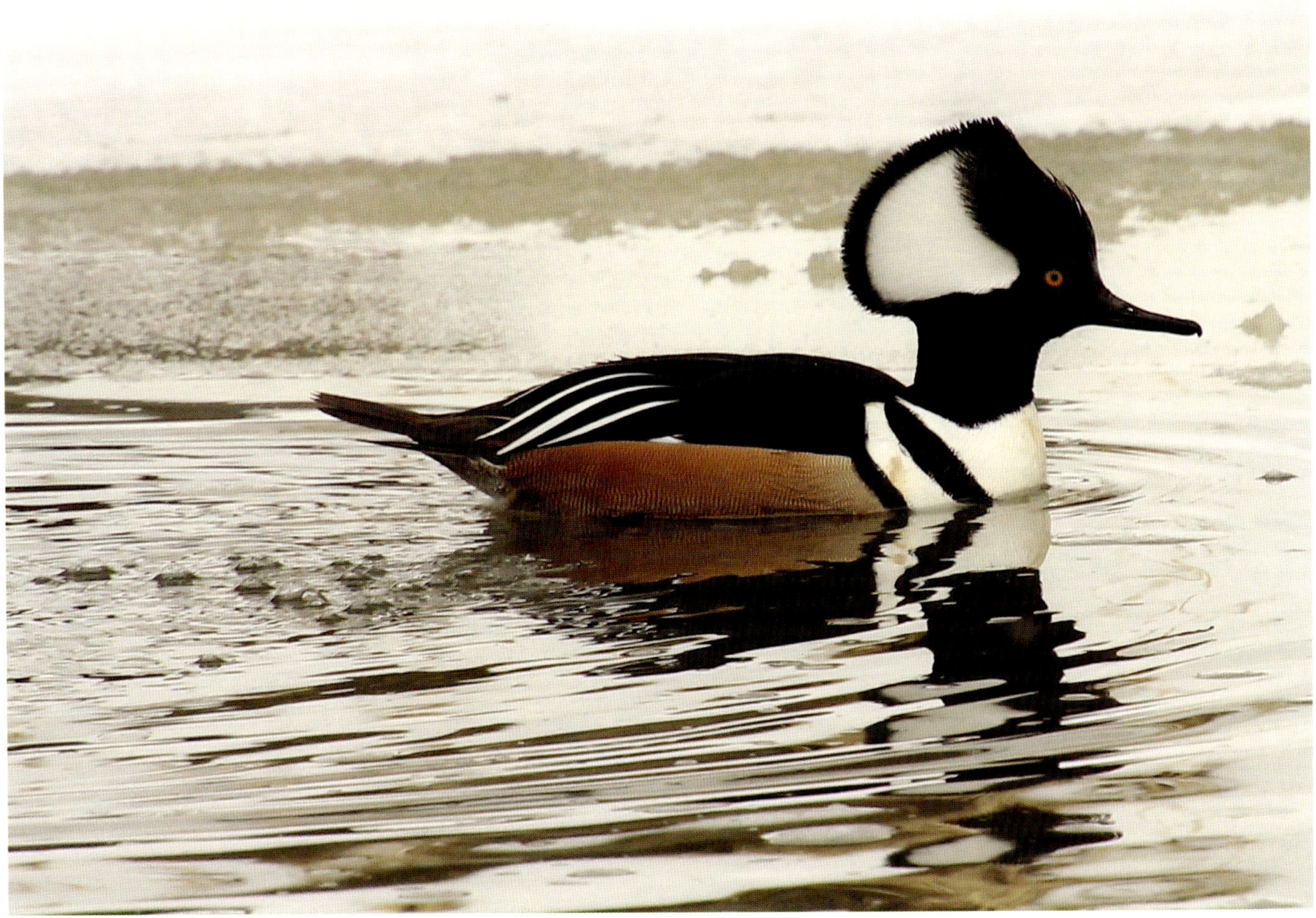

Hooded Merganser, Latah Creek, Spokane County.

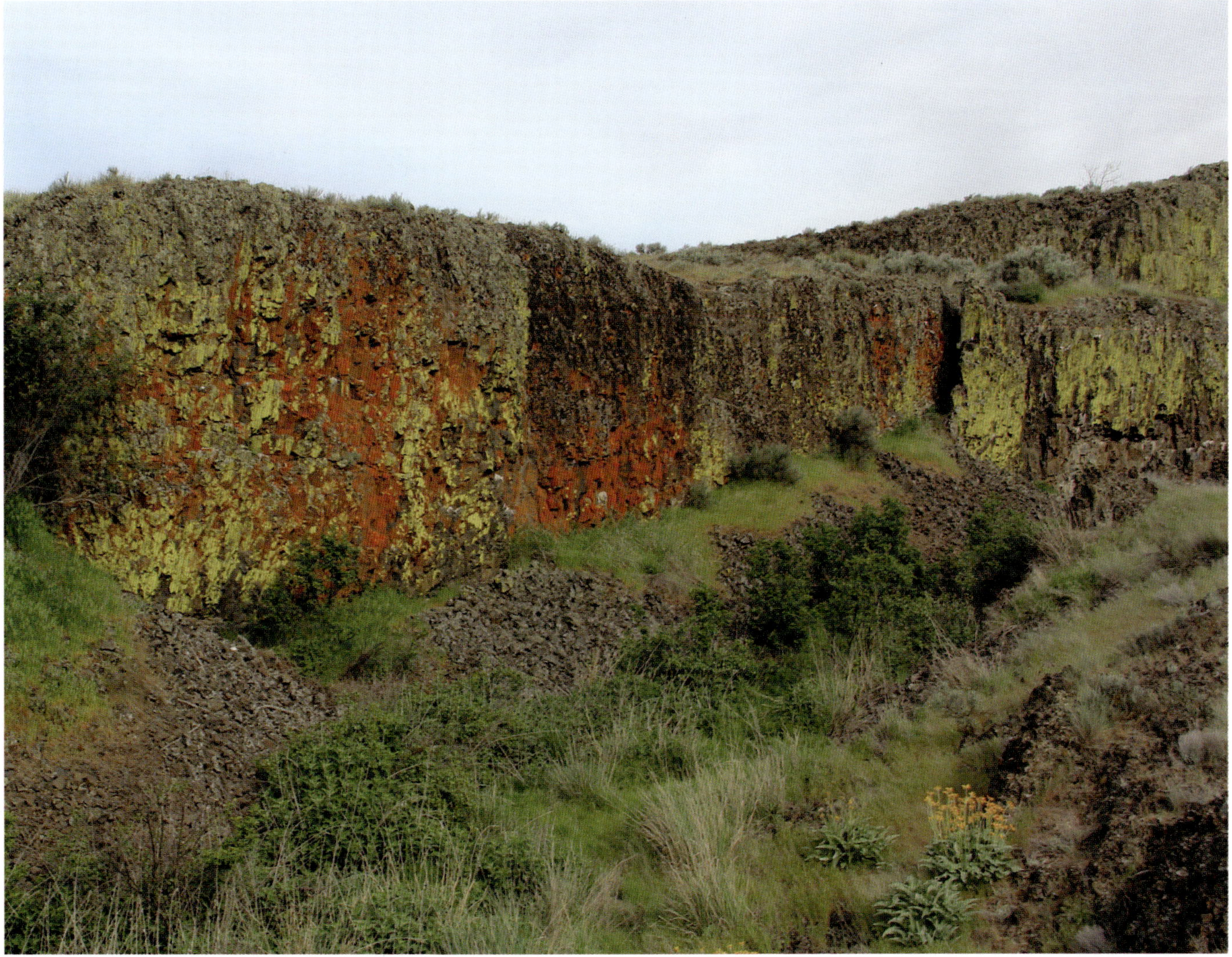

West wall, Amphitheater Crater.

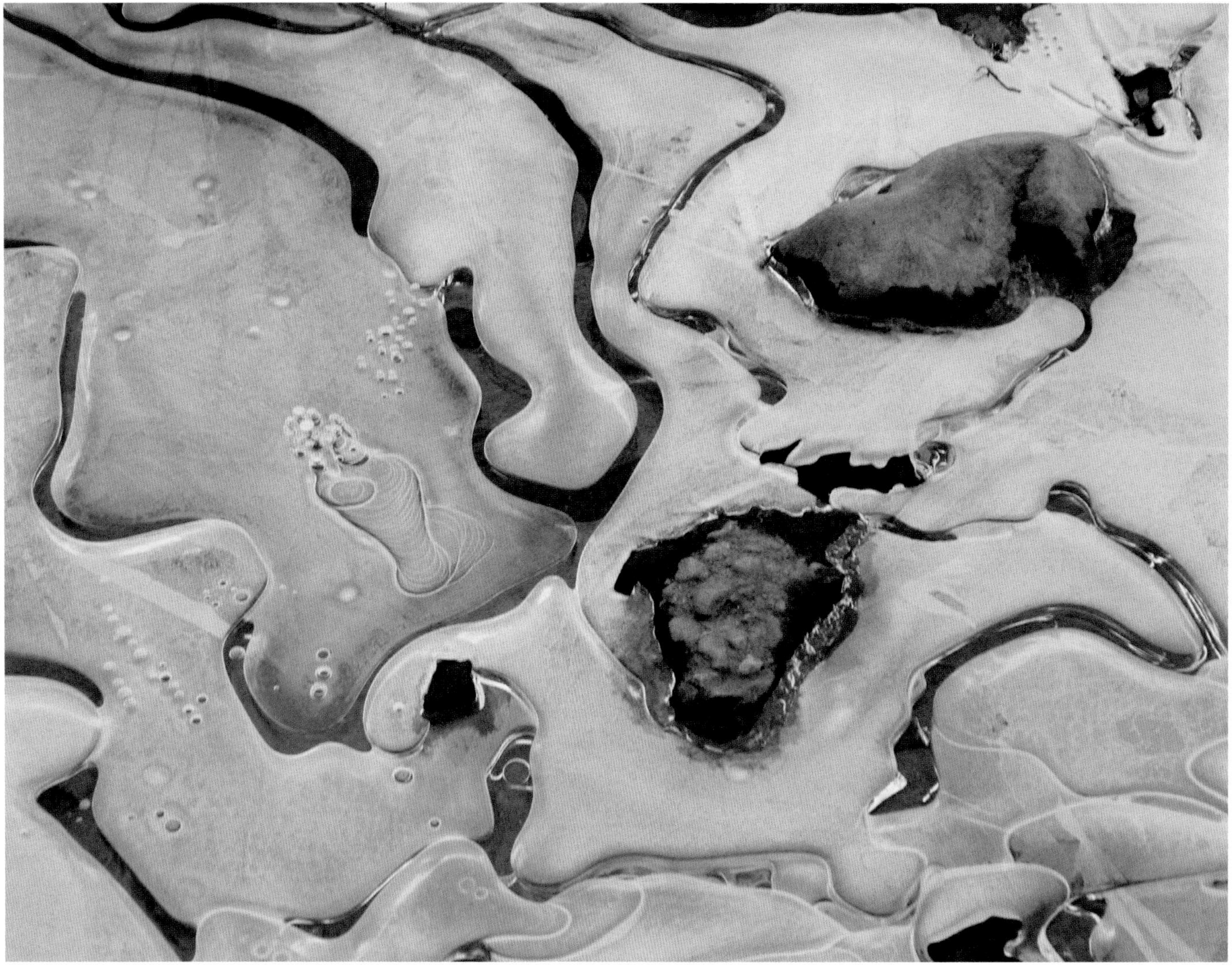

Ice at the confluence, lower Latah Creek.

EPILOGUE

The Road to Pasco

By early July the parched breeze in the gulch where I swim most summer afternoons can dry a wet dog in minutes. And yet the water has a glacial bite to it—cold enough to induce hypothermia within an hour; cold enough to force expletives from the mouths of nice people.

Clues to the *force majeure* that delivers the frigid water are everywhere in sight but difficult to grasp. Thirty-five miles to the east, comfortably lukewarm lake water stocks the Spokane River's headwaters at Coeur d'Alene, Idaho. It's just that a staggering amount of the warm water leaks through the porous riverbed long before the Spokane reaches Spokane. As the dwindling river approaches the city, the vanishing lake water is replaced by bitterly cold aquifer water—groundwater that flows through a long, deep trough of Lake Missoula outwash cobbles. The outwash aquifer starts near the southern shoreline of Lake Pend Oreille (fifty miles to the northeast) and flows all the way to West Spokane, where I get to swim in a river that would otherwise be a mere trickle by midsummer.

The icy plunges are invigorating. So is the natural beauty on the shoreline and beneath the surface. The restoration of the middle Spokane River is an environmental success story that framed the World's Fair Spokane hosted in 1974. Apart from the rafters, kayakers, tubers, and paddleboarders with whom I trade hellos, I also get to swim above vast schools of fish and beneath osprey, bank swallows, kingbirds, and bald eagles.

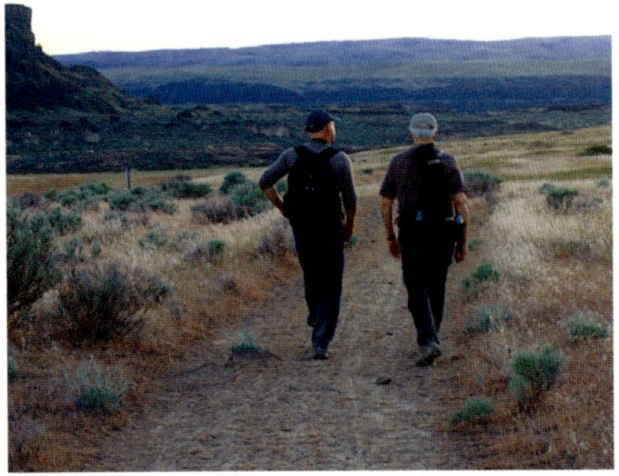

Old friends, end of the day, Potholes Coulee (*Mary Kay McCollum photograph*).

In the shallows, sunlight illuminates broad shoals of colorful cobbles and larger rocks that the ice age floods left behind. The scale of the deposits is astounding. My swimming hole—known as the Big Eddy—is bounded to the west by a dune-shaped ridge that reaches hundreds of feet above the river. It is a massive flood bar, piled all the way to the top with Missoula flood cobbles, gracefully sculpted in the same way that the floods tapered Bretz hills in the scablands to the south. Beyond the ridge is the steep rimrock of Wanapum basalt that the floodwaters and their ice-rafted cargo overflowed, dropping huge chunks of granite and argillite on the flood-scoured plains beyond.

It is plausible that ancient Native Americans witnessed one or more of the Bretz floods. A quarter century ago, a nearly complete skeleton of "The Ancient One"—originally dubbed "Kennewick Man"—was discovered near Pasco. Subsequent testing dated the bones to be more than 8,000 years old. An artifact at the Marmes (pronounced MAR-mus) Rock-shelter on the Palouse River a few miles below Palouse Falls has been dated at more than 11,000 years old. There is ample oral history of the great floods among inland Northwest tribes.

I try to imagine what a Bretz flood would have looked and sounded like as it roared into Northrup Canyon or future townsites such as Lamont or Othello. For the animals and any humans in harm's way, it must have been a terrifying demise. My dad taught me to swim even before he taught me to tie my shoes. I imagine the breathless futility of trying to bodysurf a Missoula flood wave.

The consuming problem of a broken heart is that it feels as I imagine a Missoula flood wave would have felt. It crushes your spirit and washes away hope. At times I felt as though I were trapped in a phone booth, drowning in the most intimate pain imaginable. Getting out under the sky, to breathe away from my life in town, was the best I could do. I have dear people in my life who will console me in hard times. I just didn't want them to do so.

My friend Larry Shook and I met in 1980, the year Mount Saint Helens erupted. A few months later he hired me to write for *Spokane Magazine*, a superb, independent city magazine that he and his wife published. It was the seed of a remarkable partnership that kept bringing us together for most of three decades. It was always intense work, culminating in a lengthy, white-collar crime story that won several national journalism prizes. It was also tumultuous and costly. Our investors abandoned us. Neither of our marriages survived.

Larry is a decade older than I am, and that is the main reason he went to Vietnam and I didn't. He was a door gunner in helicopters, survived being shot down more than once, and lived through the ghastly carnage of the Tet Offensive in 1968. I only learned about his service in Vietnam because he alluded to it in a column he wrote about his unease with the Armed Forces Torchlight Parade that is the centerpiece of Spokane's annual Lilac Festival. Otherwise, he was not inclined to discuss the topic.

In the fall of 2012, he was driving at dusk on a rural highway in northern Idaho. A mature deer stepped out onto the roadway. Although he and his companion were unhurt, the bloody collision triggered his long-suppressed psychological trauma from Vietnam. Months later, on Memorial Day 2013, I was with him when he delivered a guest sermon at Spokane's Unitarian Universalist Church. He shared his story and spoke vividly and emotionally about the psychological injuries that so many surviving warriors bring home with them.

It was a difficult time for both of us, in no small part because we were both reeling from the loss of our marriages. I was stubborn about my desire to walk alone, both for the soul-searching and the patience that the photography often required. But when Larry asked to join me on at least some of my hikes, I couldn't refuse. And that's how we wound up in the Drumheller the day the cranes sang down to us, and chatting up horseback riders (and their horses) near the Escure Ranch in remote Whitman County, and in Potholes Coulee on a long day that ended beneath the northern lights. We've learned well, over the past forty years, how to carry each other, and my dear friend also carries my tripod.

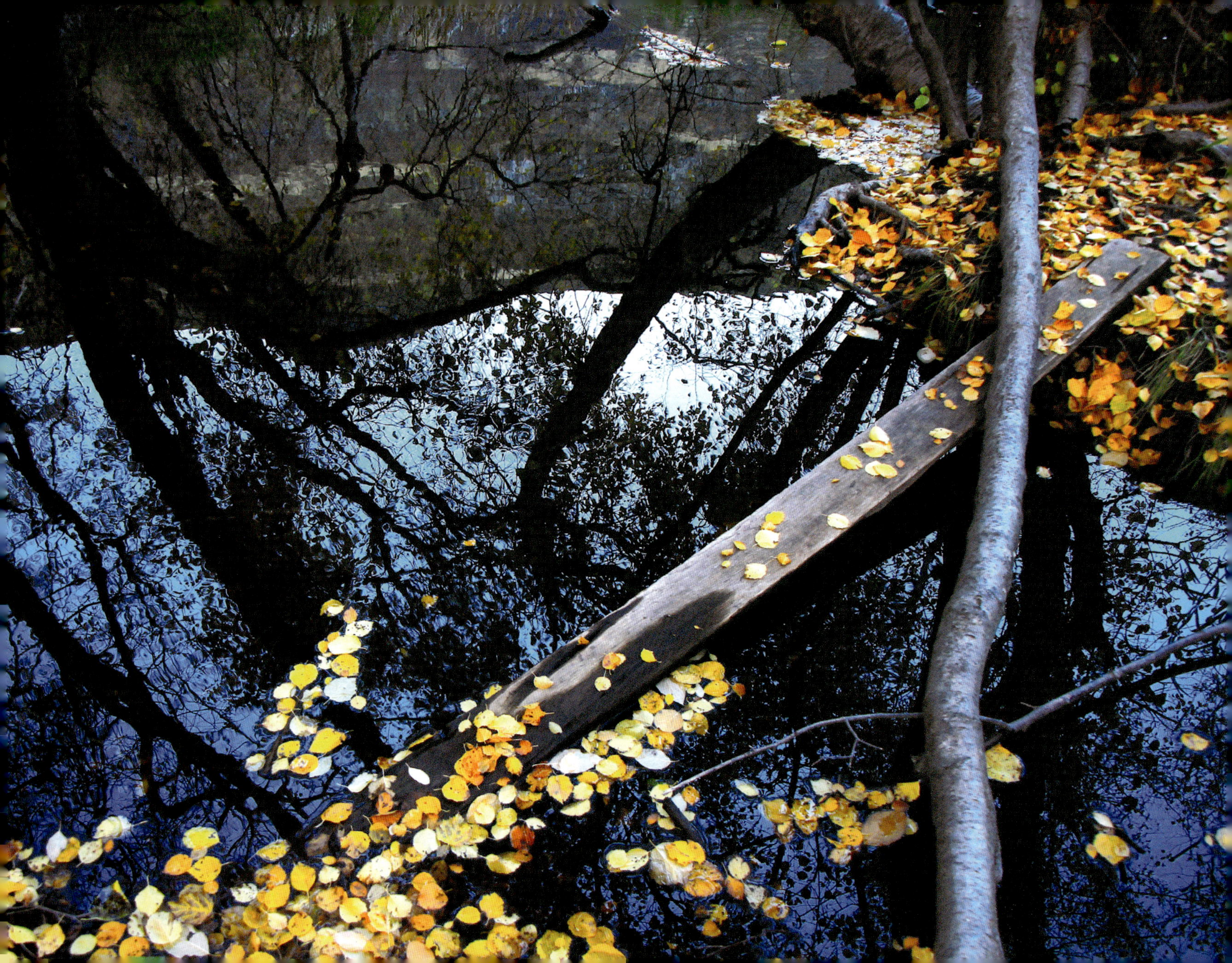

In *The Solace of Open Spaces*, Gretel Ehrlich's account of her retreat into the wilds of Wyoming after the death of her lover, she writes that "cascading water equates loss followed by loss, a momentum of things falling in the direction of death, then life."

That's an apt metaphor for the solace of walking the scablands' floodscapes. With time our faces are furrowed by the storms of our lives and the salt of our tears, just as the Bretz hills were simultaneously ravaged by the floods and transformed into tear-shaped dunes on the landscape, their prow-like faces still looking in the direction of the hurt.

As I write this paragraph it is nineteen days since Joan died. My mother was eighty-six and fulfilled her wish of passing at home. Three days before she took her last breath, I offered her water through a straw and she drank nearly a whole cup. I cheerfully congratulated her and, hearing my voice, she searched for my eyes, smiled playfully, and rolled her shoulders like a passing wave, as though she were dancing the samba.

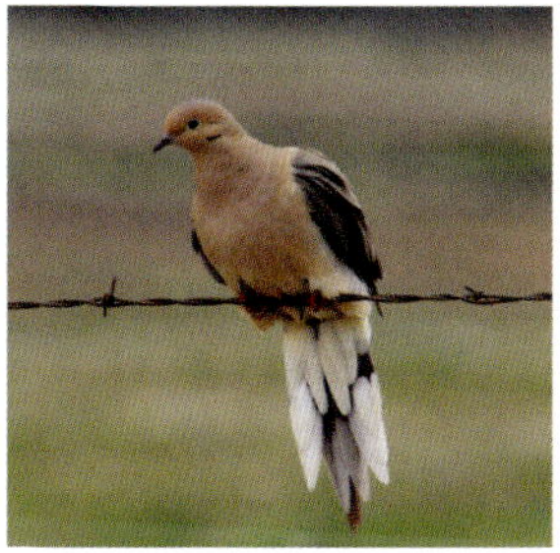

In her last three years I slept on a couch beside her most nights. As her dementia progressed she would more frequently wake up in the darker hours of the morning, asking me to take her home, to Pasco, so she could be with her father. She worried that Gil would be worried about her.

"God's country," I remember her saying. She's home now.

Left: Meadow Creek, Dry Falls State Park.

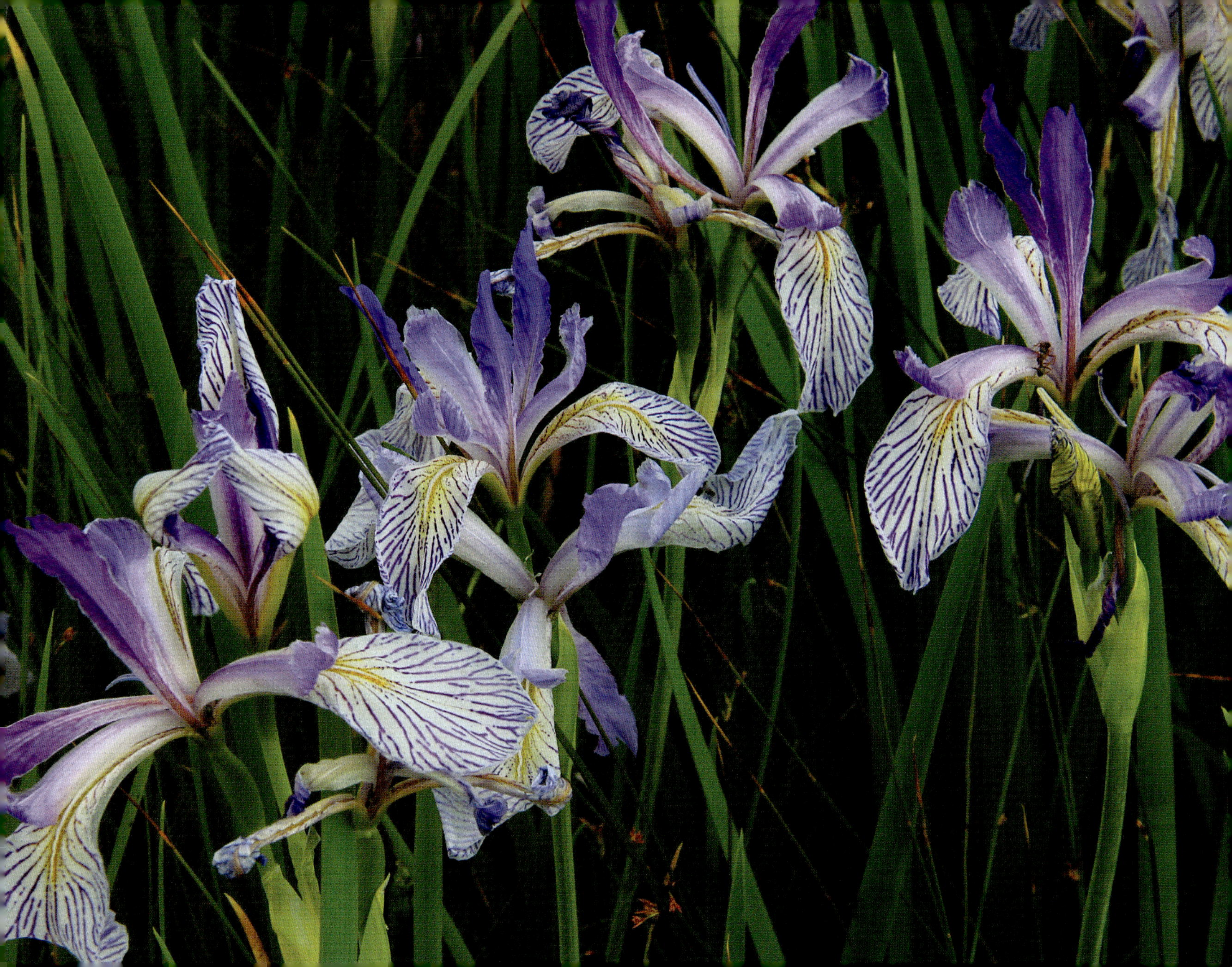

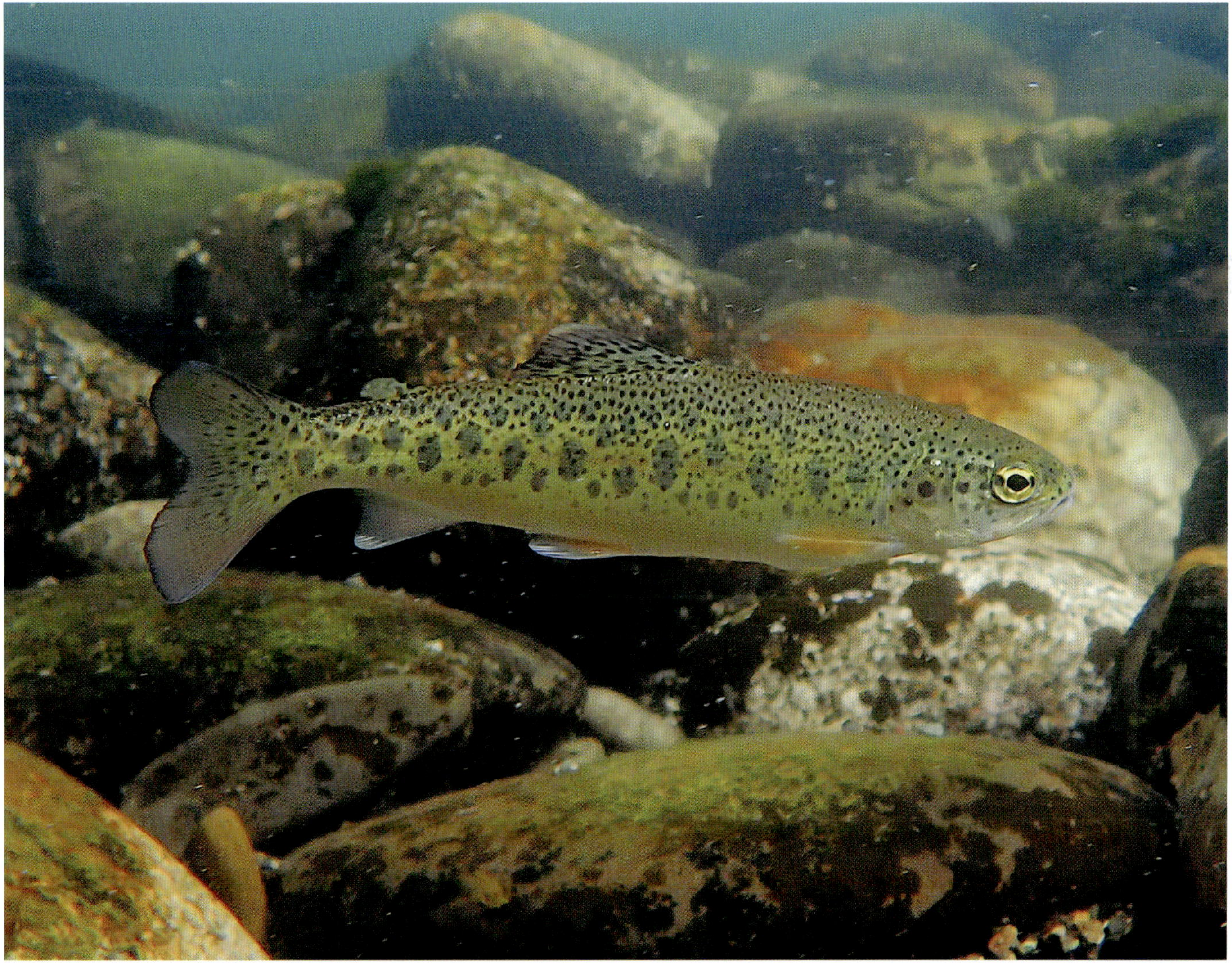

Left: Rocky Mountain irises, Lake Creek Coulee. *Right:* Juvenile native Redband trout, Spokane River.

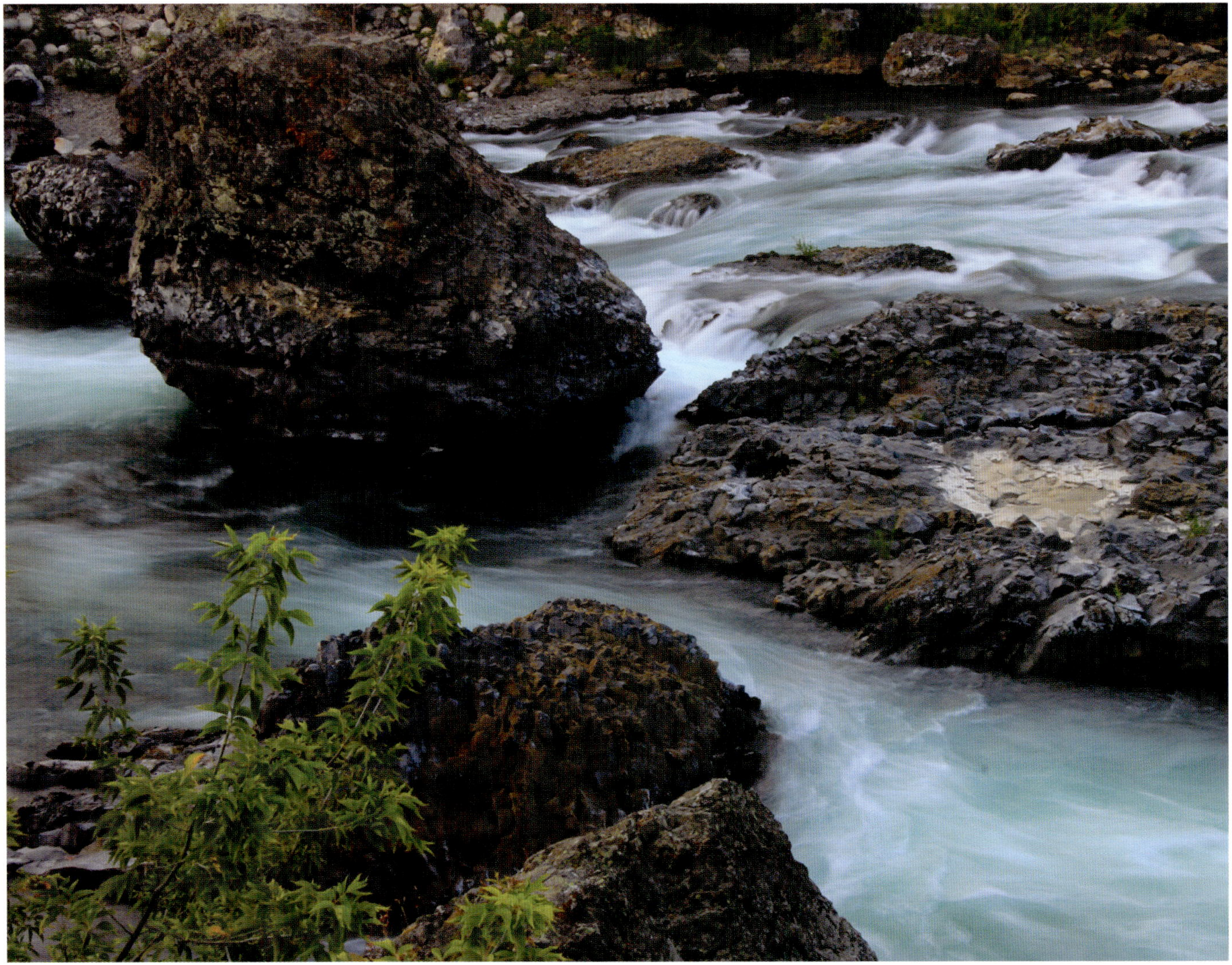

Left: Devil's toenail, Spokane River at Riverside State Park. *Right:* The Feathers in February, Frenchman Coulee.

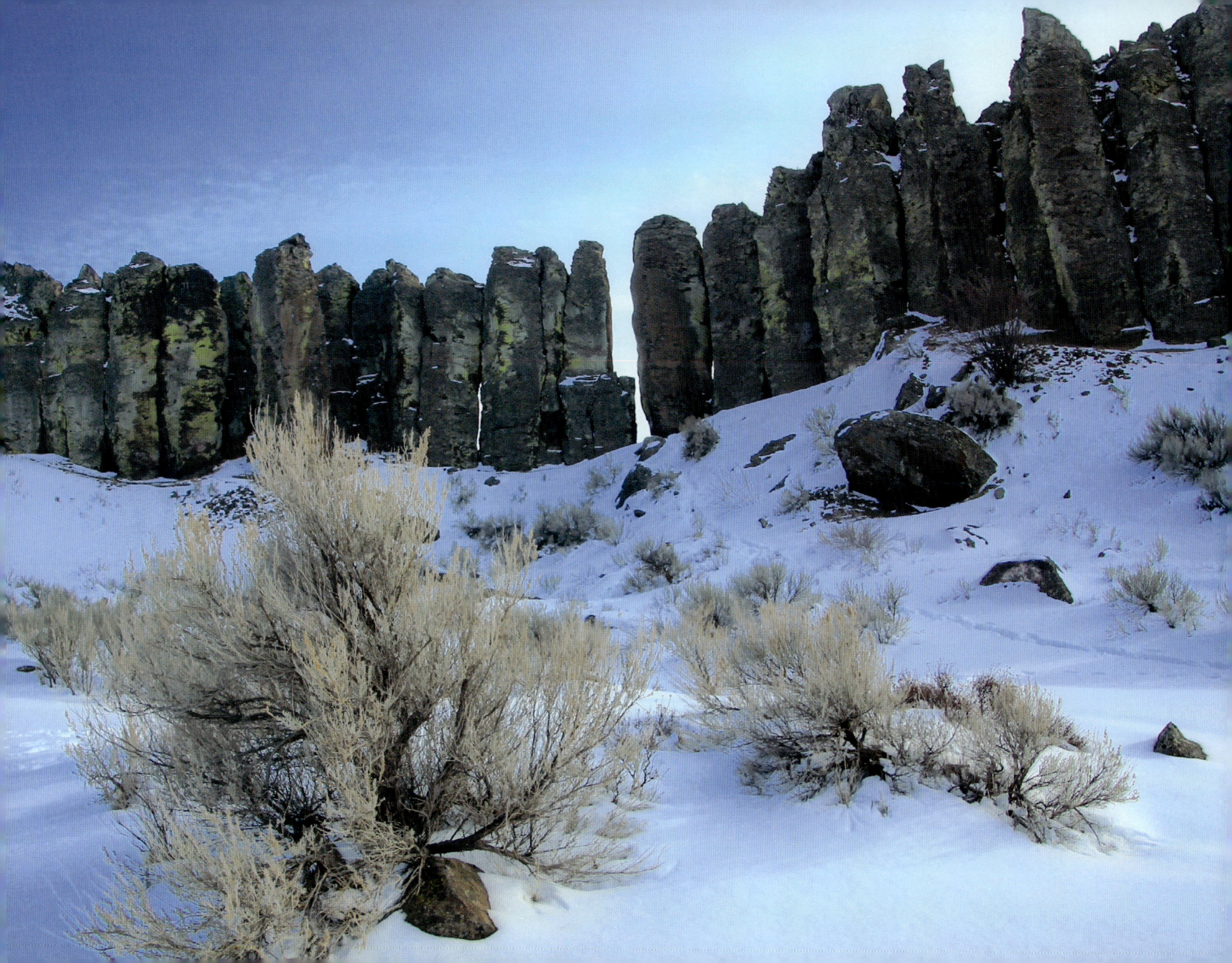

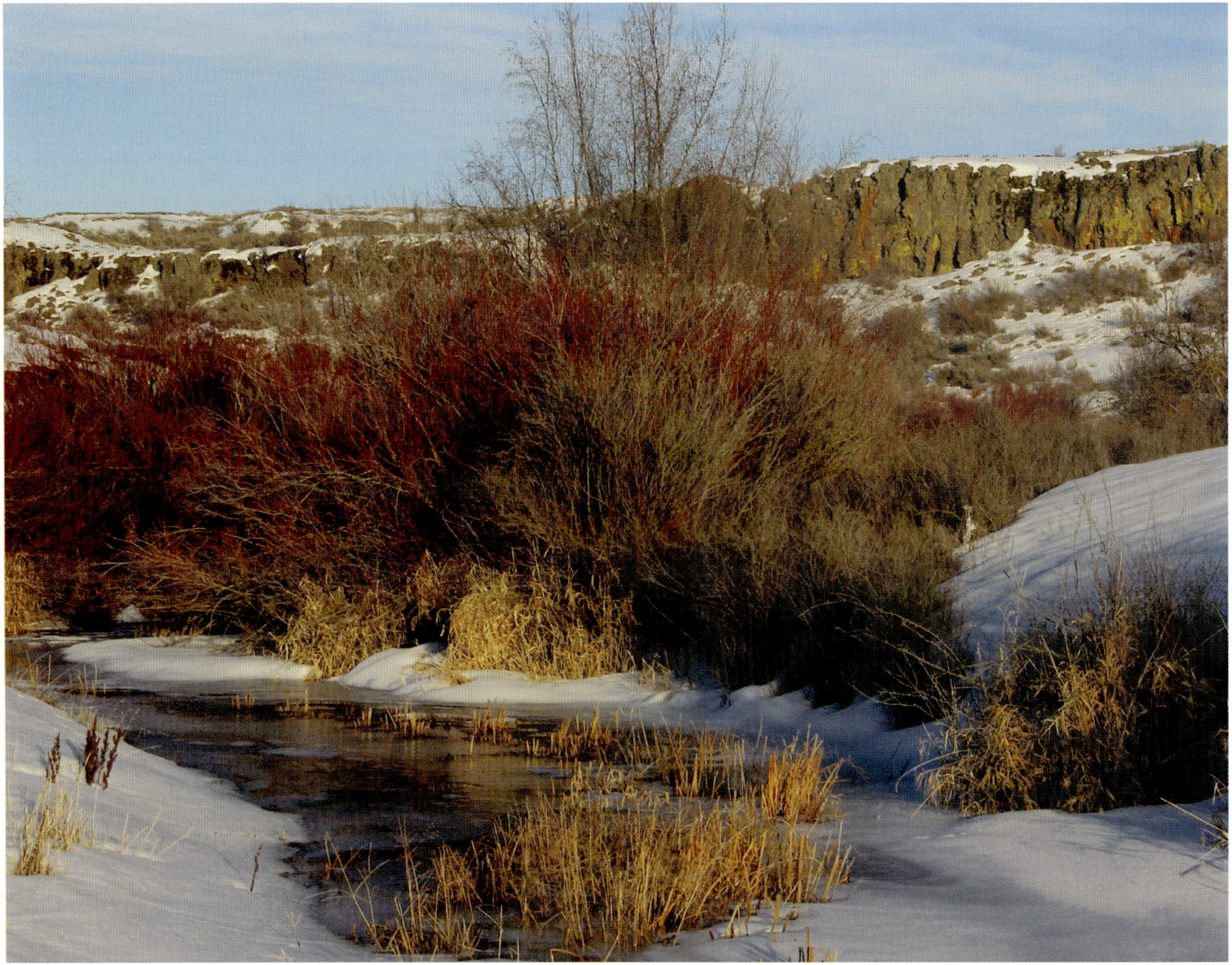

Left: Winter range, Lake Creek Coulee. *Right:* Northern Flicker.

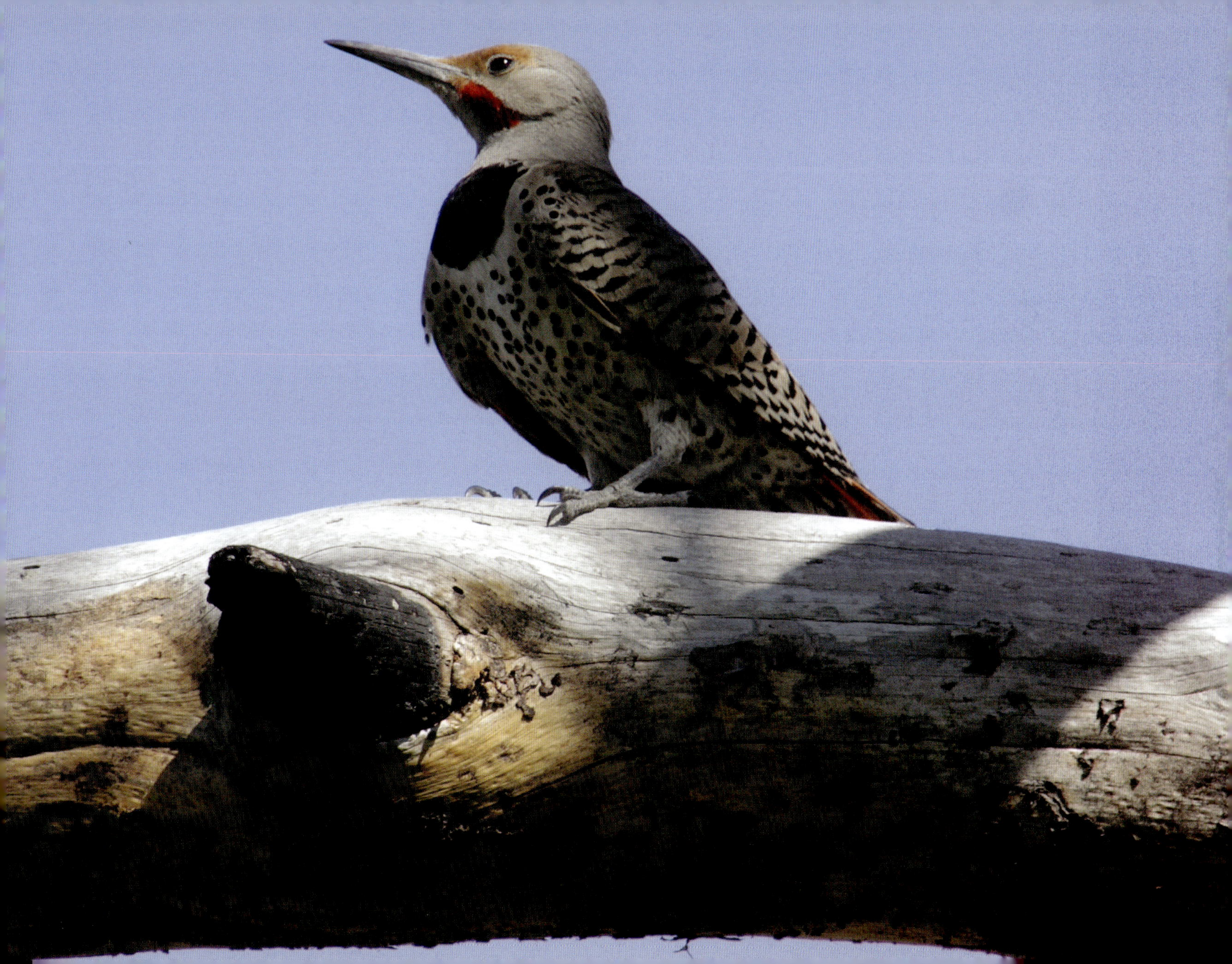

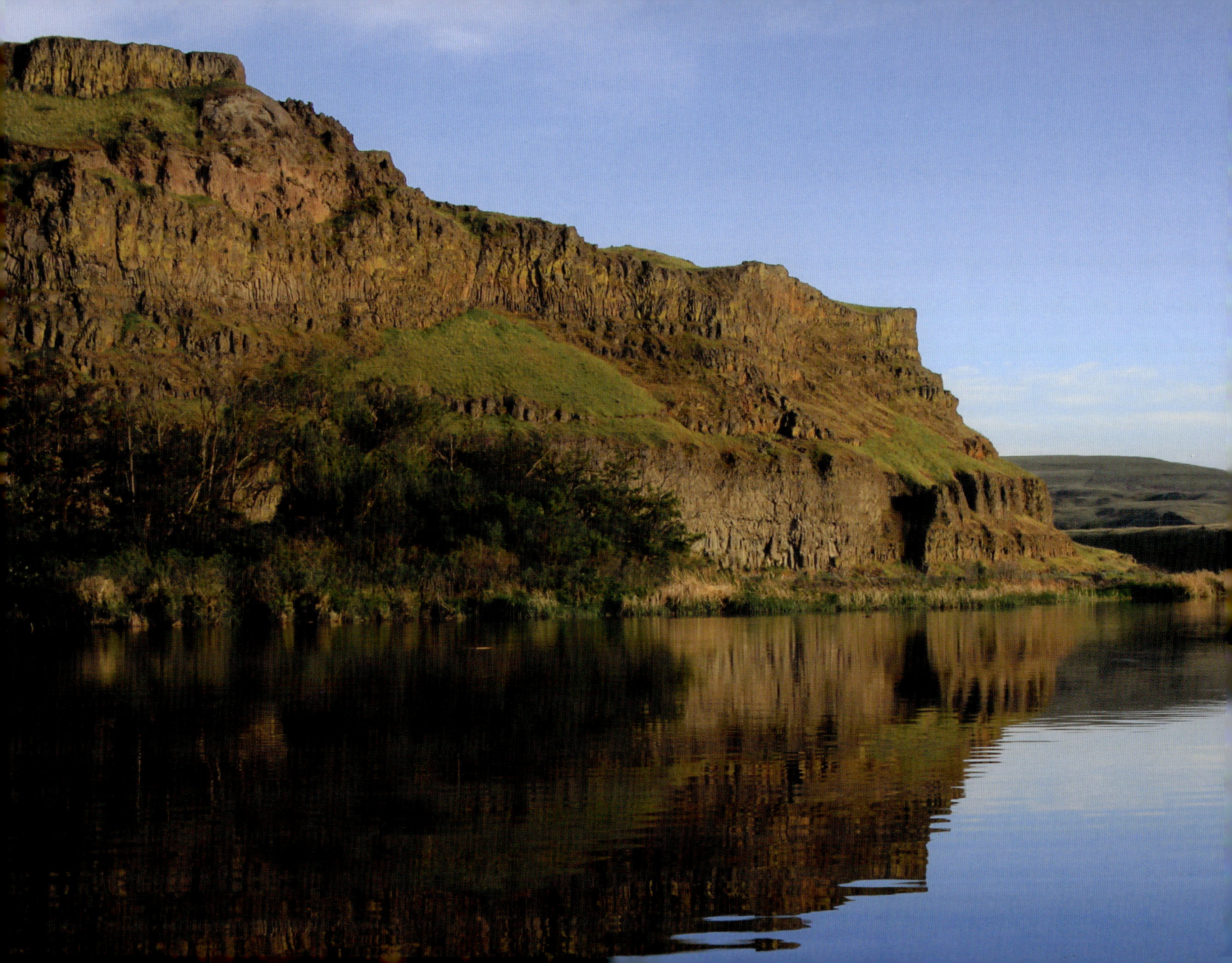

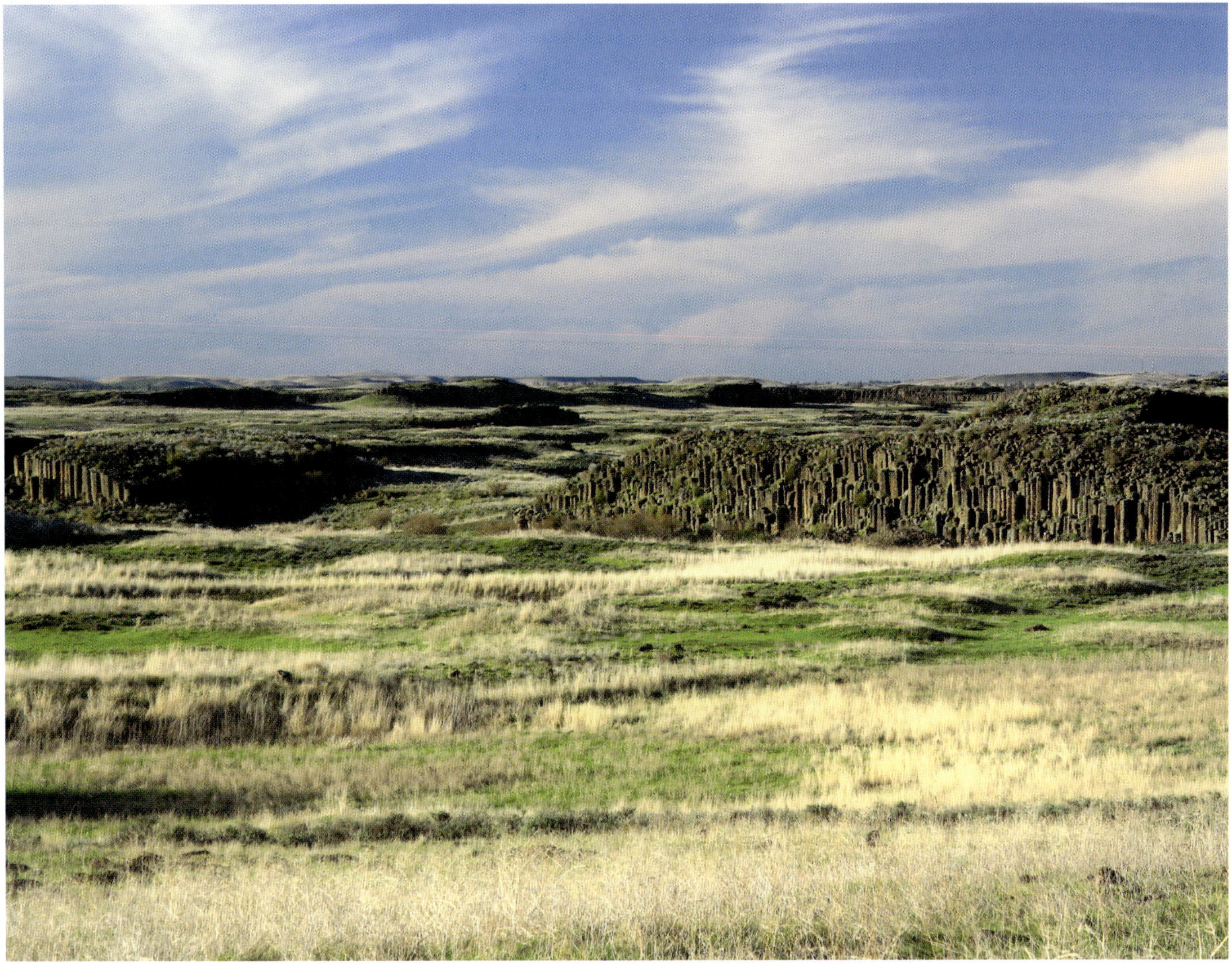

Left: Palouse River at Lyons Ferry. *Right:* Scabland terrain, northwestern Whitman County.

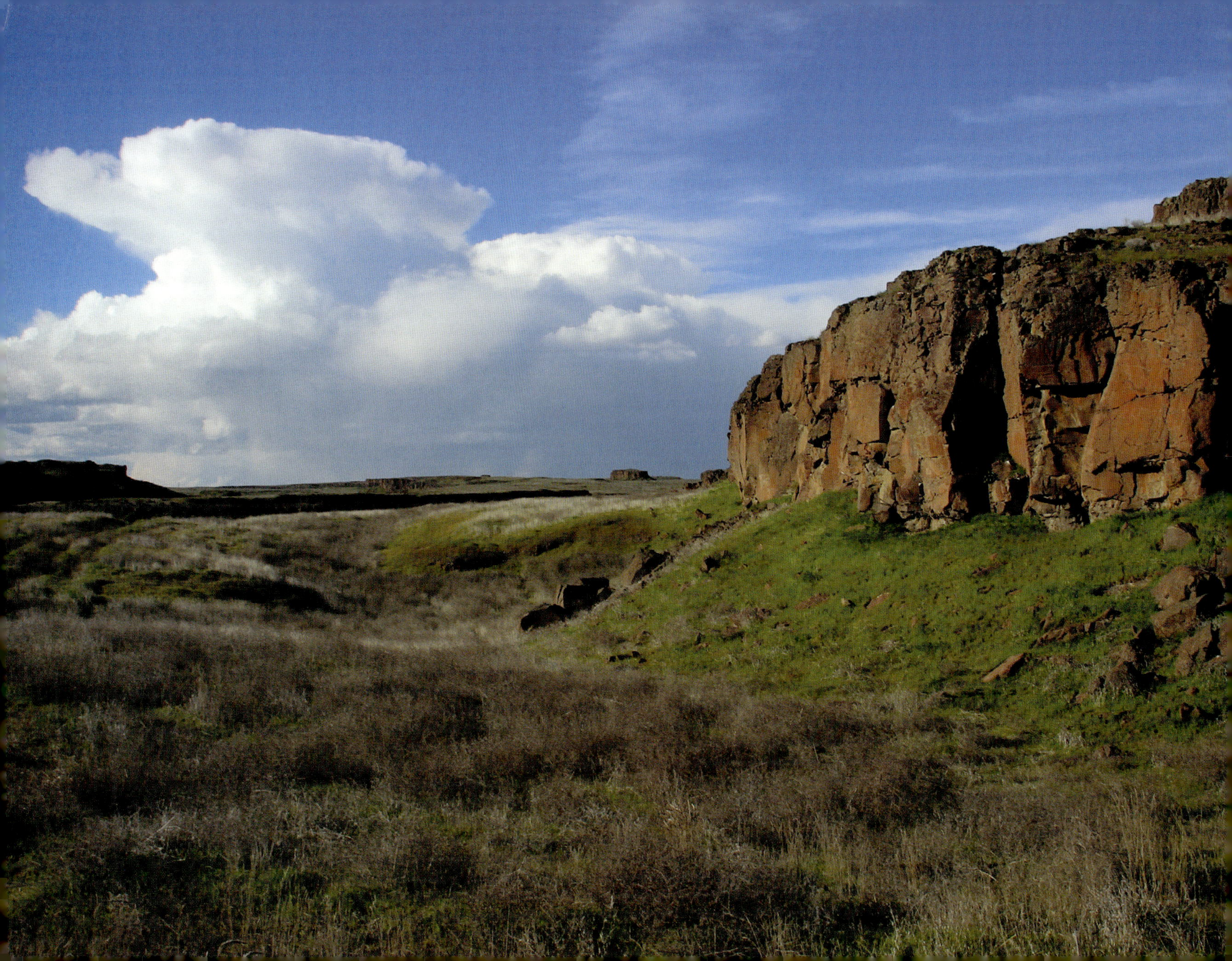

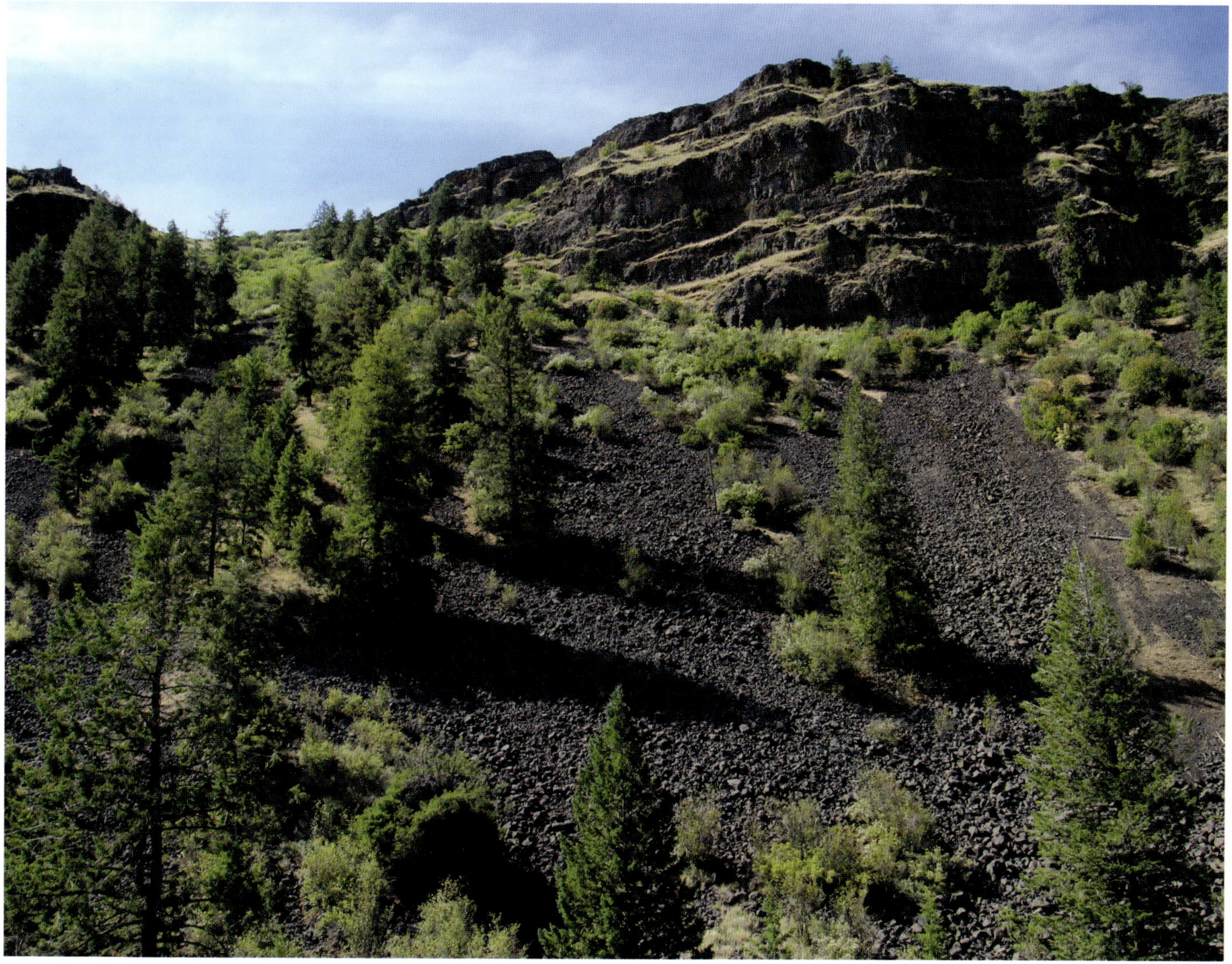

Left: Summer thunderhead, Drumheller Channels. *Right:* Highlands near Grand Coulee.

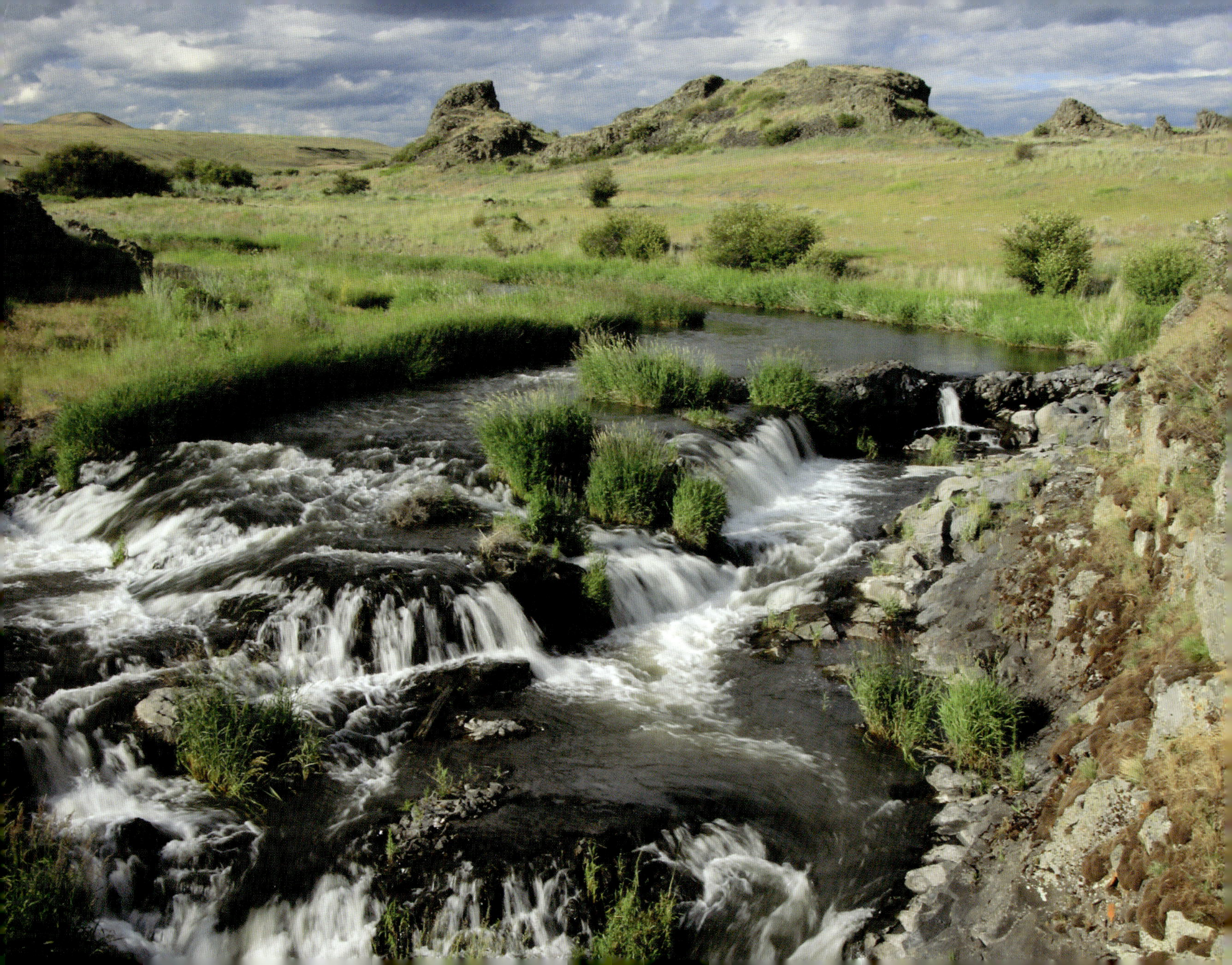

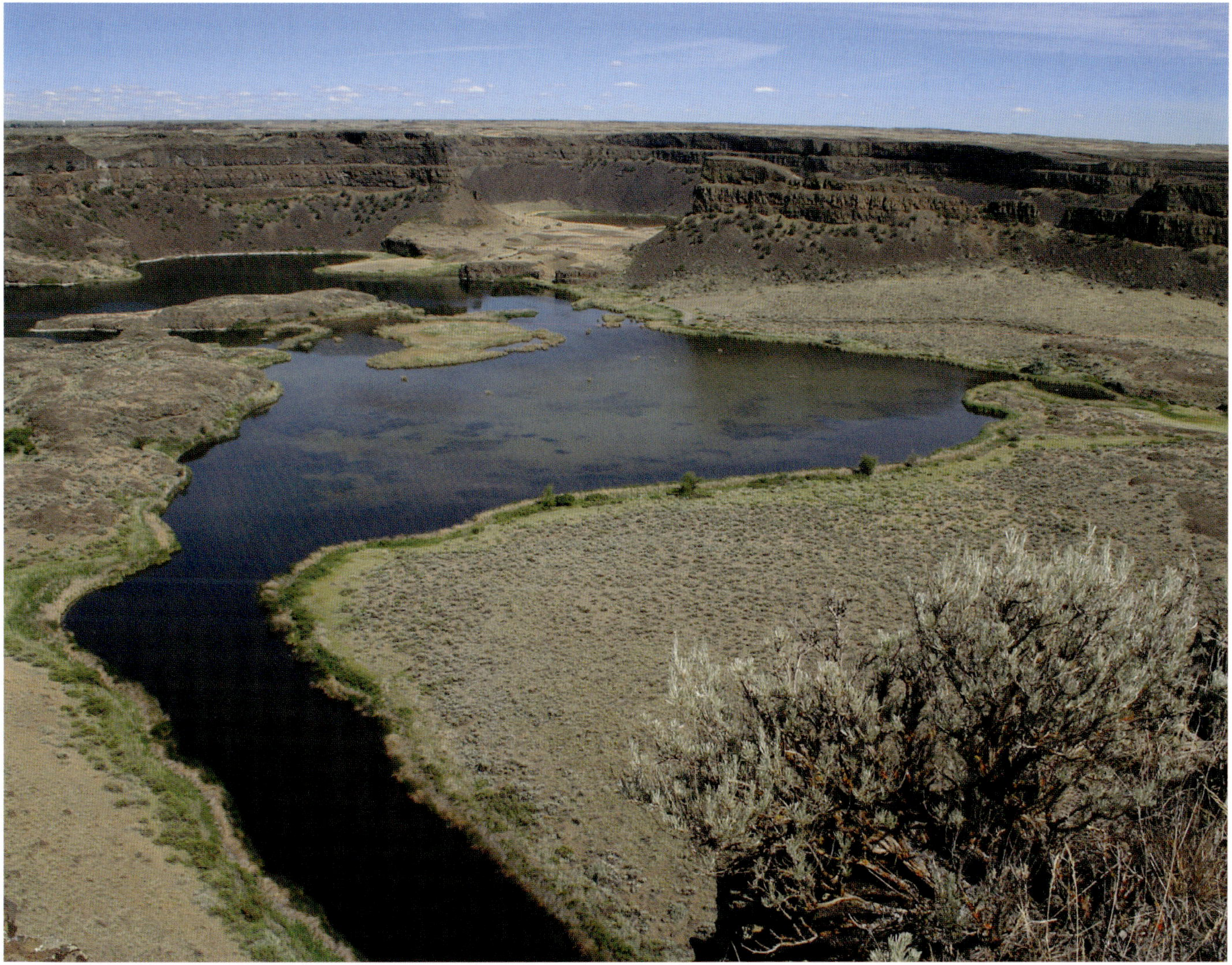

Left: Dry Falls, lower Grand Coulee. *Right*: Rock Creek, south of Revere.

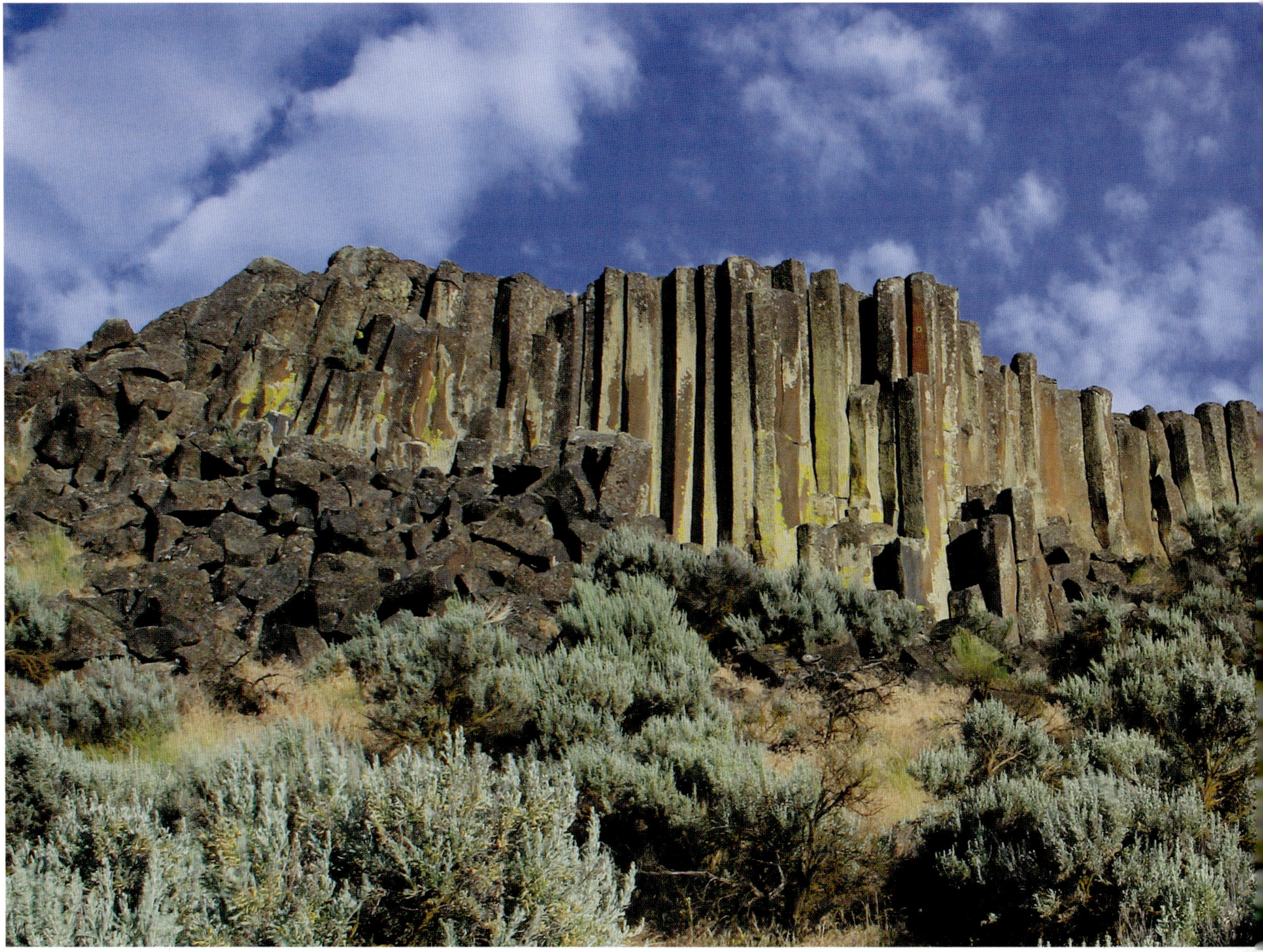

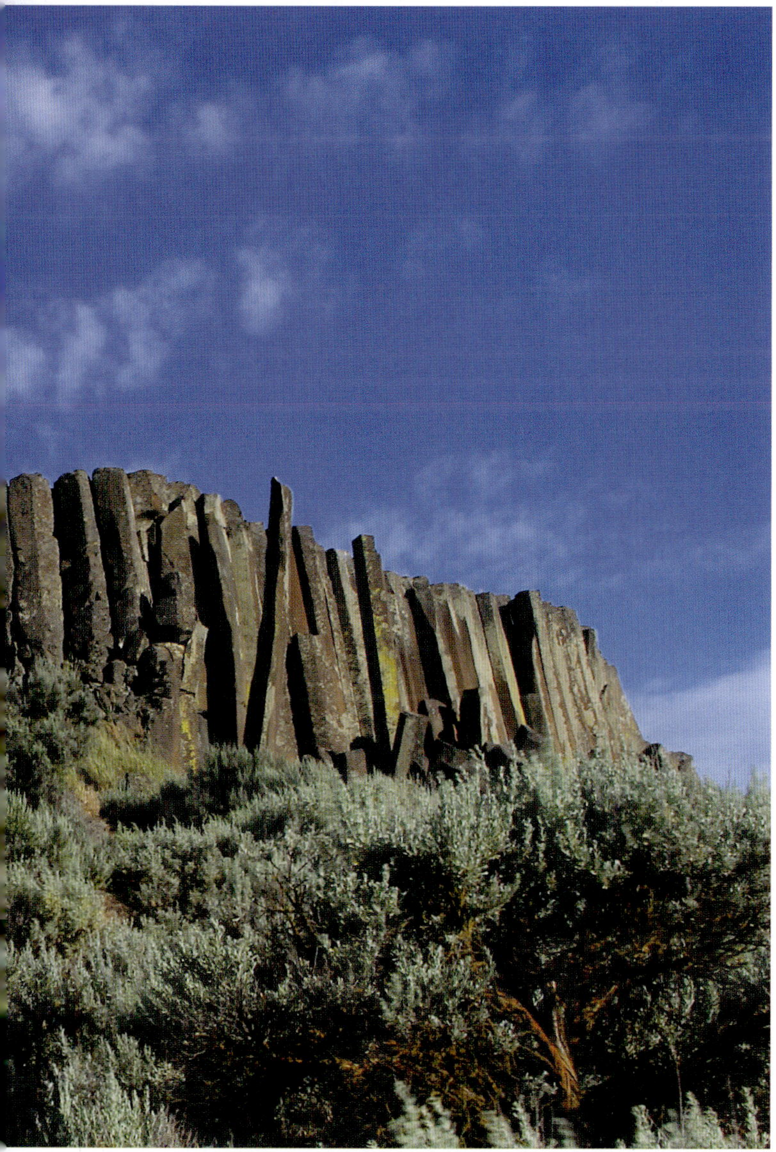

Sage and Elephant Mountain basalt palisade, north of Othello.

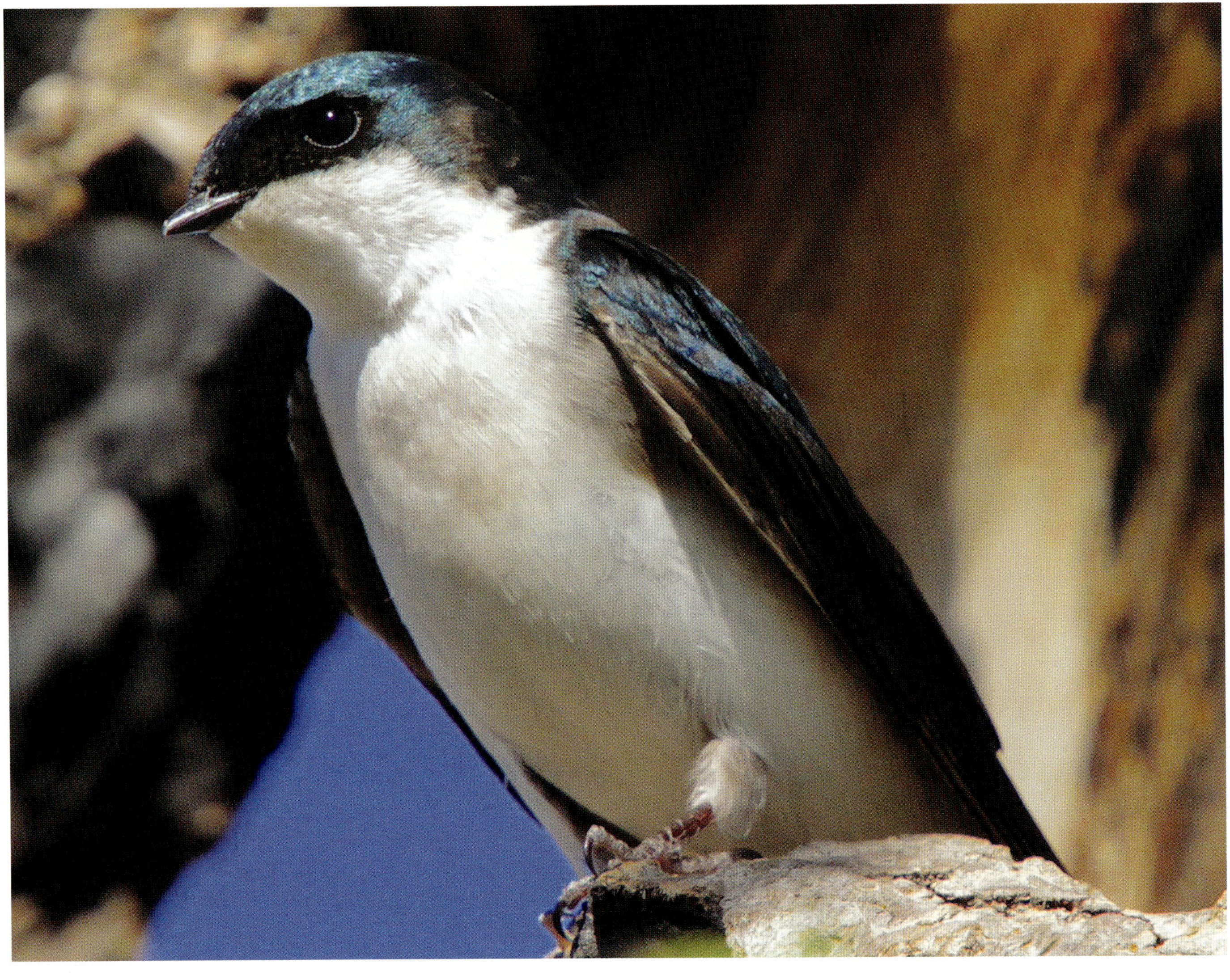

Left: Perching tree swallow. *Right:* Lone poplar near Rock Lake.

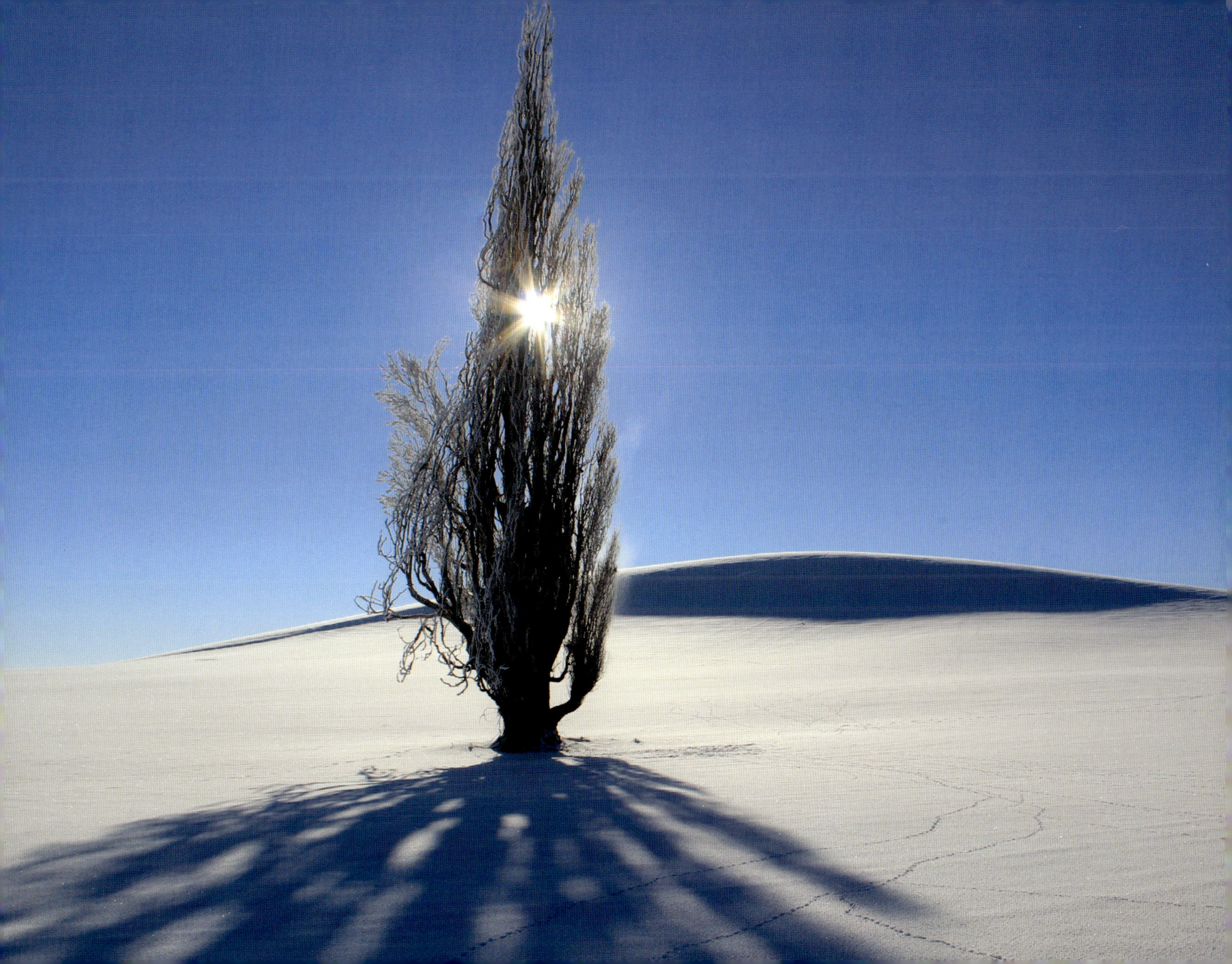

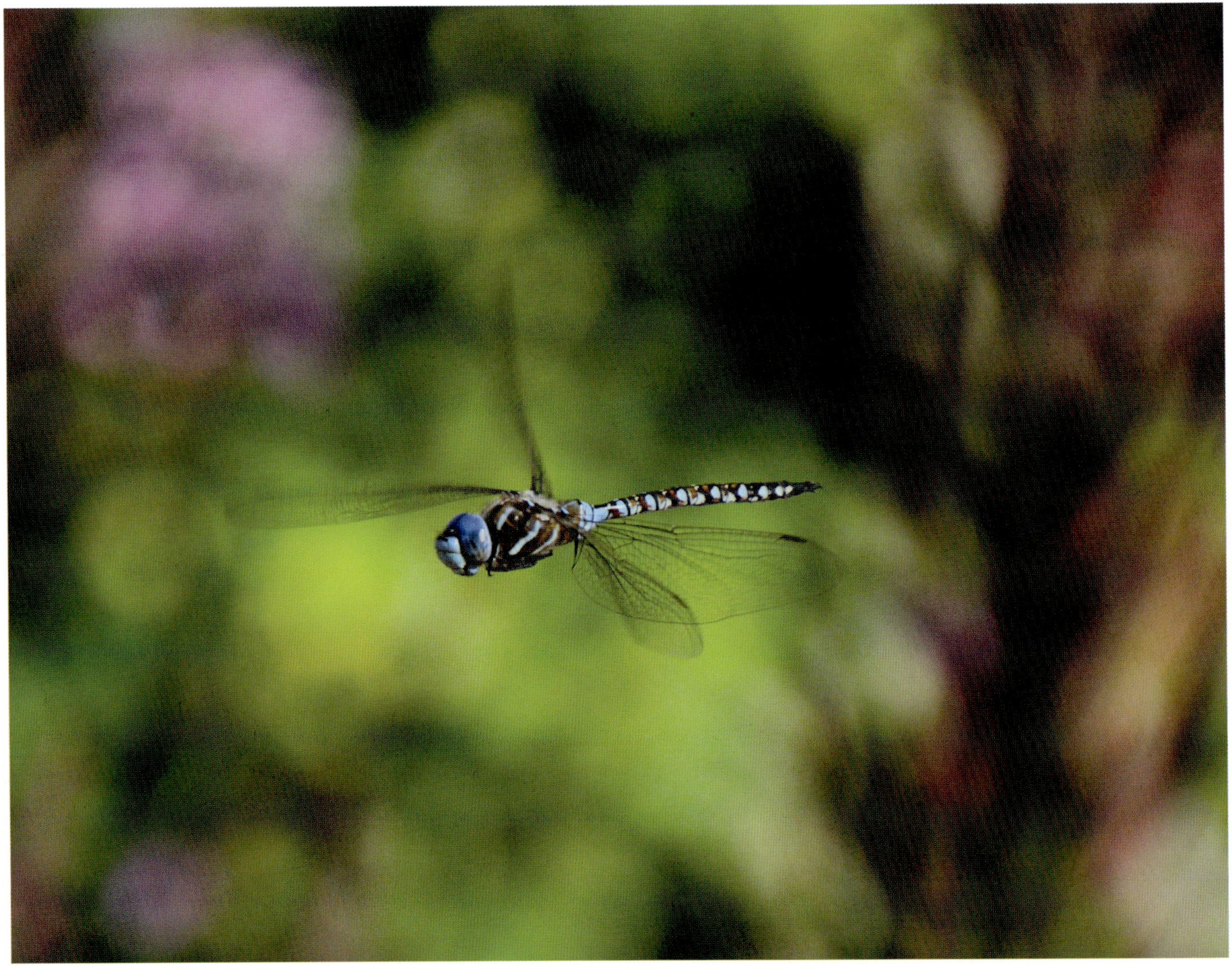

Left: Blue Darner dragonfly and fireweed at a scabland marsh. *Right:* Spokane River flowing through Grande Ronde basalt pillars, west of Spokane.

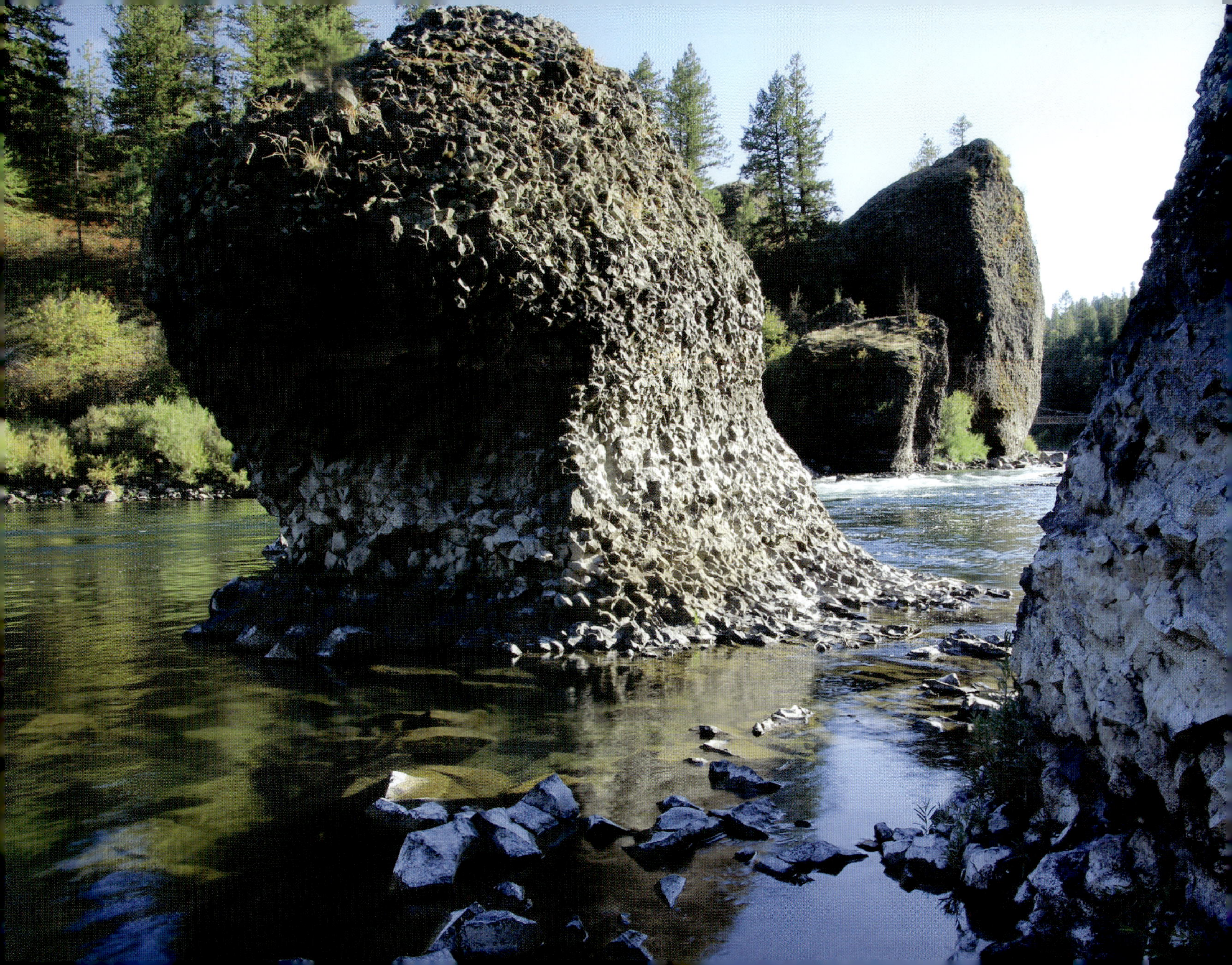

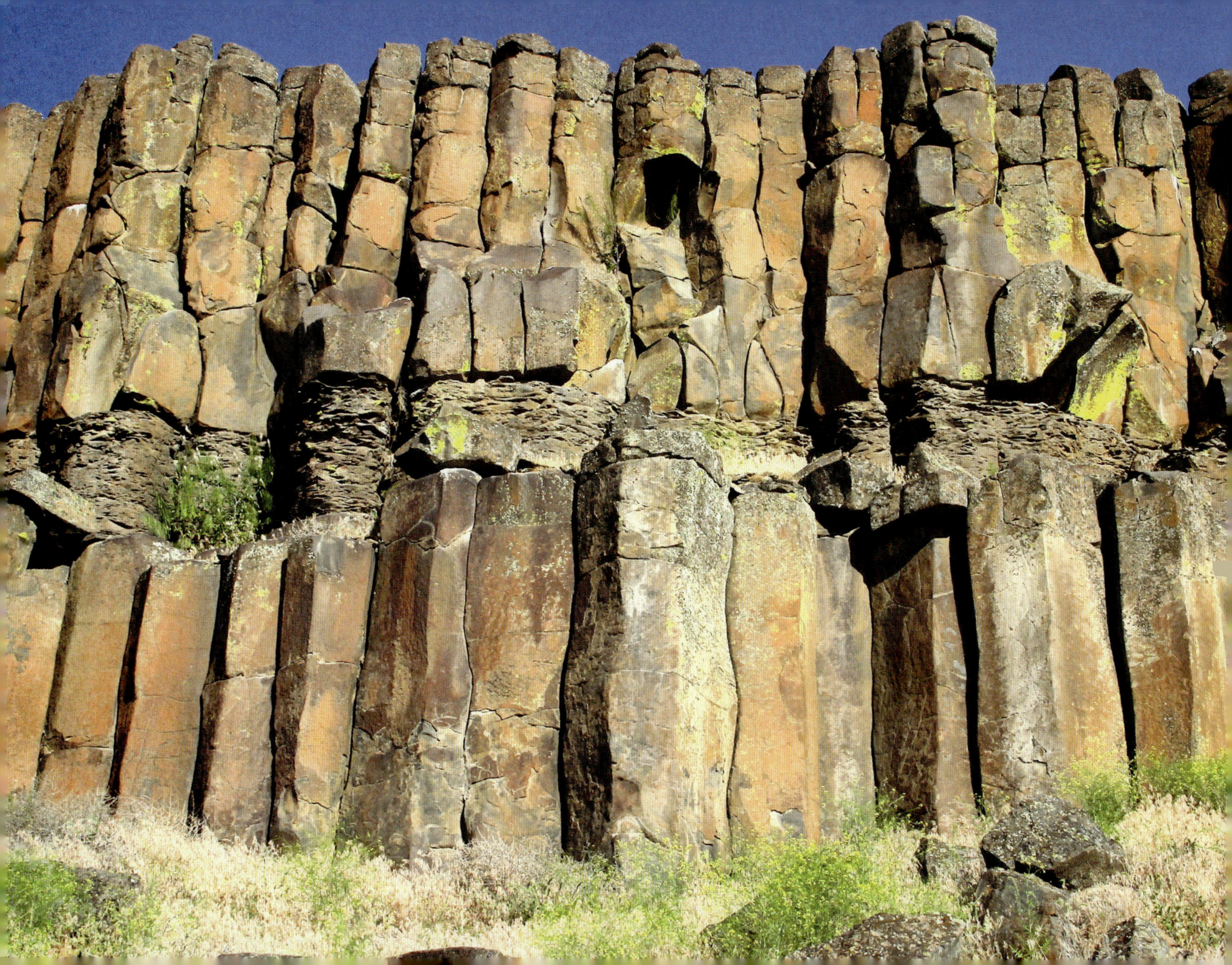

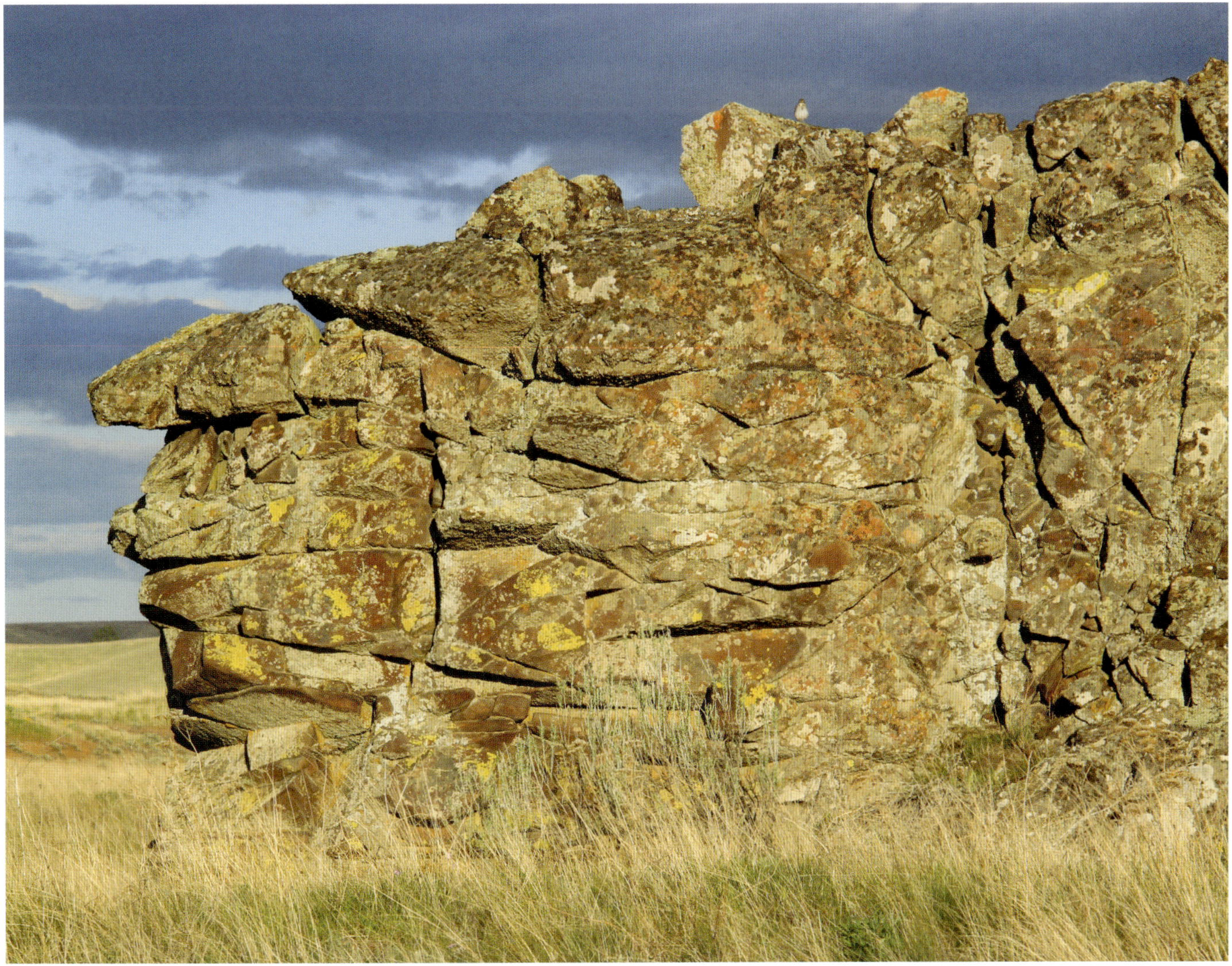

Left: Basalt columns, Drumheller Channels. *Right:* Basalt outcrop west of Lamont.

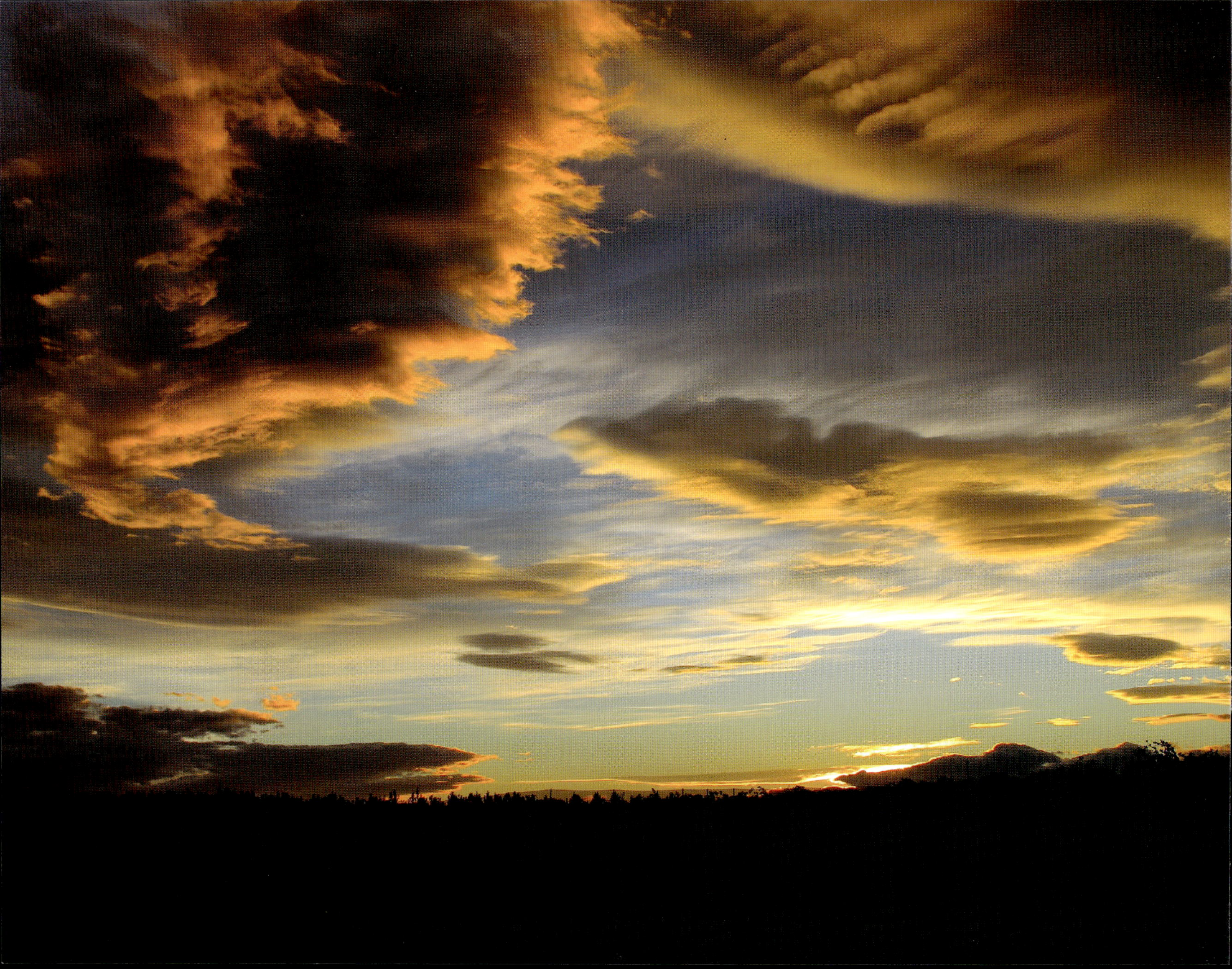

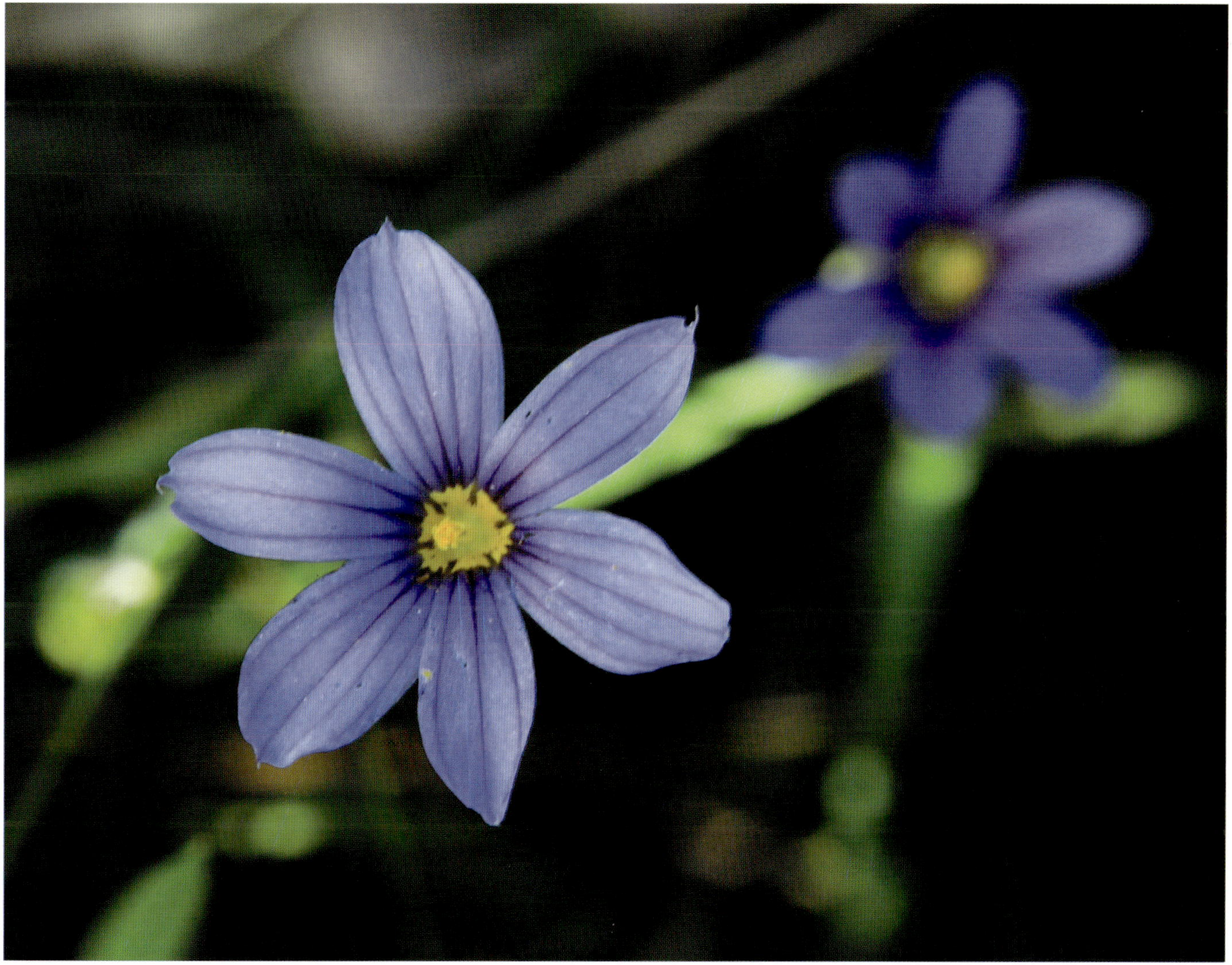

Left: Desert sunset east of Pasco. *Right:* Idaho blue-eyed grass.

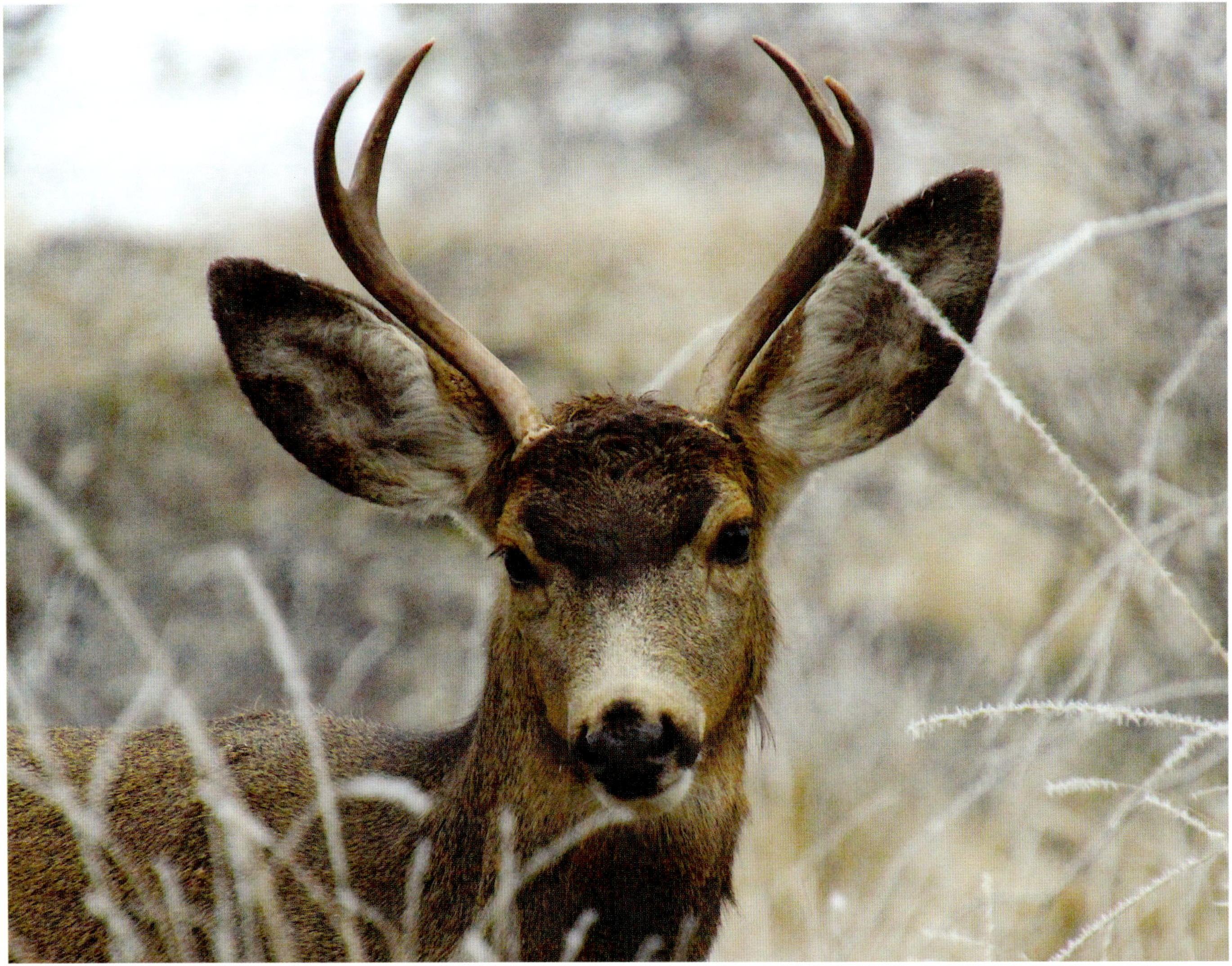

Left: Young mule deer buck in scabland meadow, west of Spokane. *Right*: Larch in the rimrock, west of Spokane.

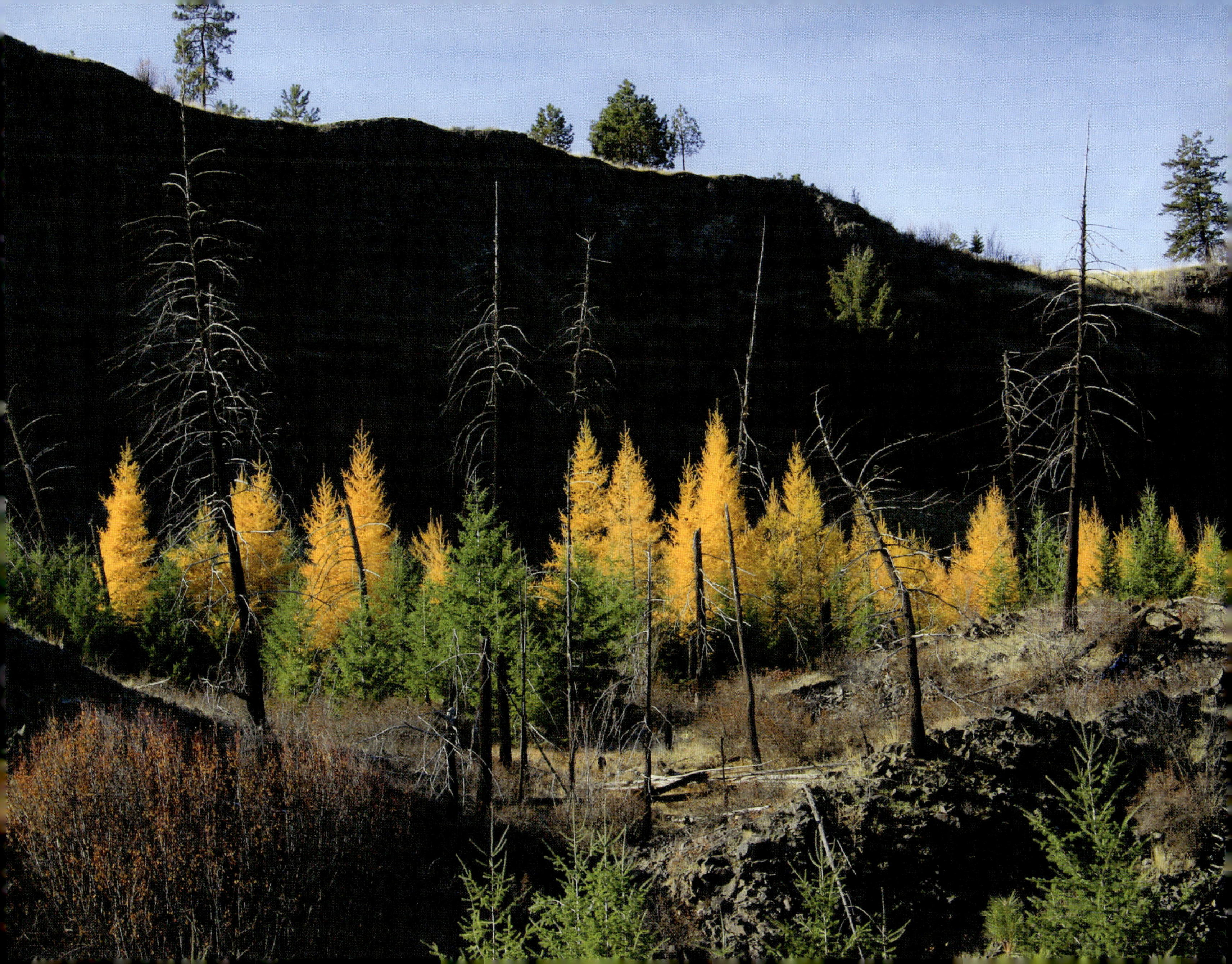

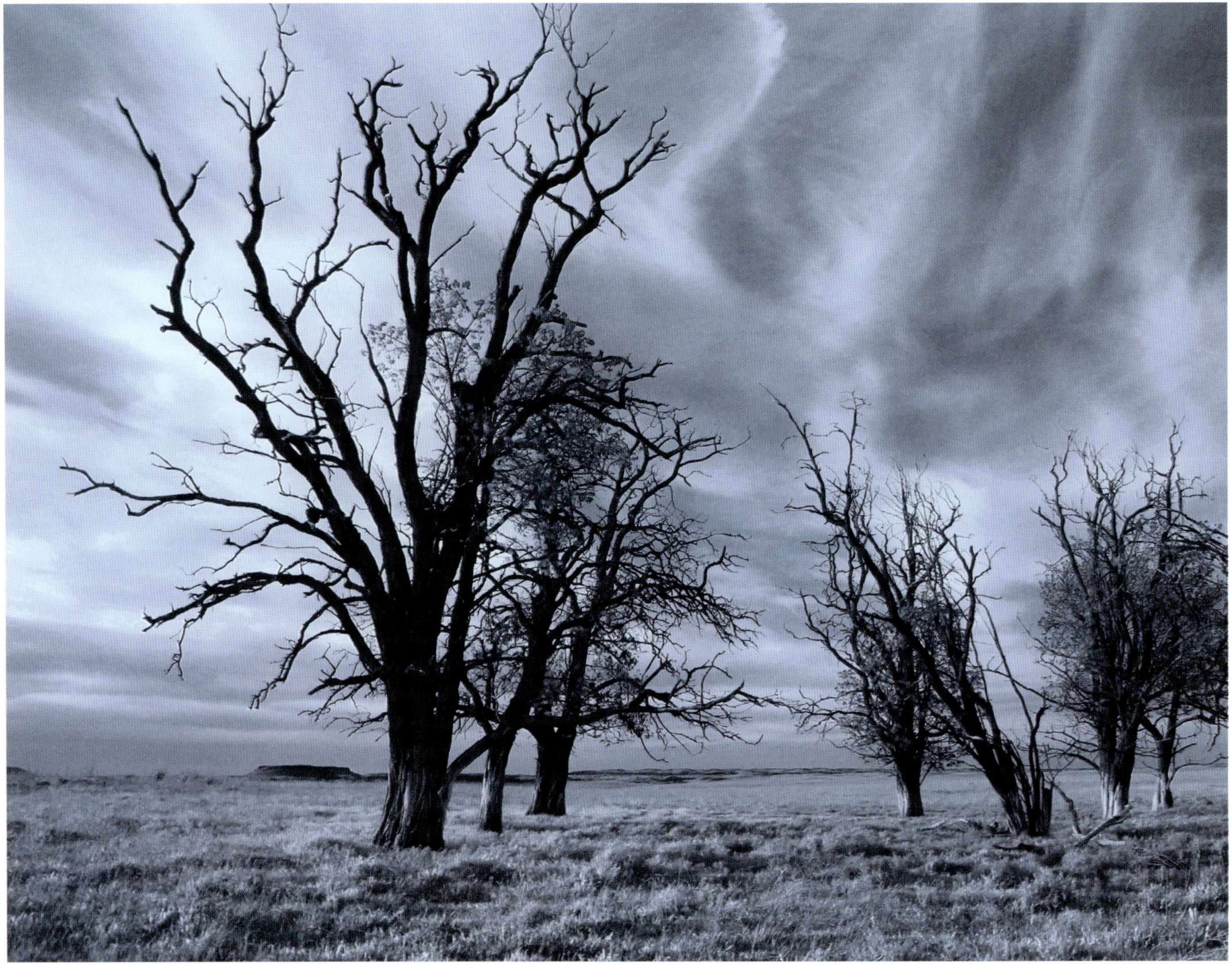

Abandoned farm site, central Adams County.

ACKNOWLEDGMENTS

Months before *Beautiful Wounds* became the working title for this book, it was the name of a large photography collection. The exhibit—complete with recorded birdsong—was headed for the spacious civic gallery in the thriving city of Moses Lake, Washington, near the epicenter of the scablands. The godmother of the exhibit (which opened in March 2020) was my friend, artist Ildikó Kalapacs, who has been a loyal supporter of my journalism for more than two decades. The enthusiastic staff at Moses Lake thought a narrative guide to the photography would help explain the exhibit and how it came about. That request became the embryo for a manuscript. It is a story that gathered weight and form with the help of two friends, Jon Gosch and Larry Shook, both talented editors. When my friend David Brookbank learned about the project in late 2019, he insisted I share the manuscript and photography with his sister Joan, who specializes in connecting writers, artists, and photographers with publishers. Joan's tenacious support and guidance for *Beautiful Wounds* has been invaluable. Special thanks to the Countryman Press editors Michael Tizzano, Isabel McCarthy, and James Jayo, and to artistic director Allison Chi, production manager Devon Zahn, and their colleagues, who worked so well during such a long and tumultuous year to help bring this work to light.

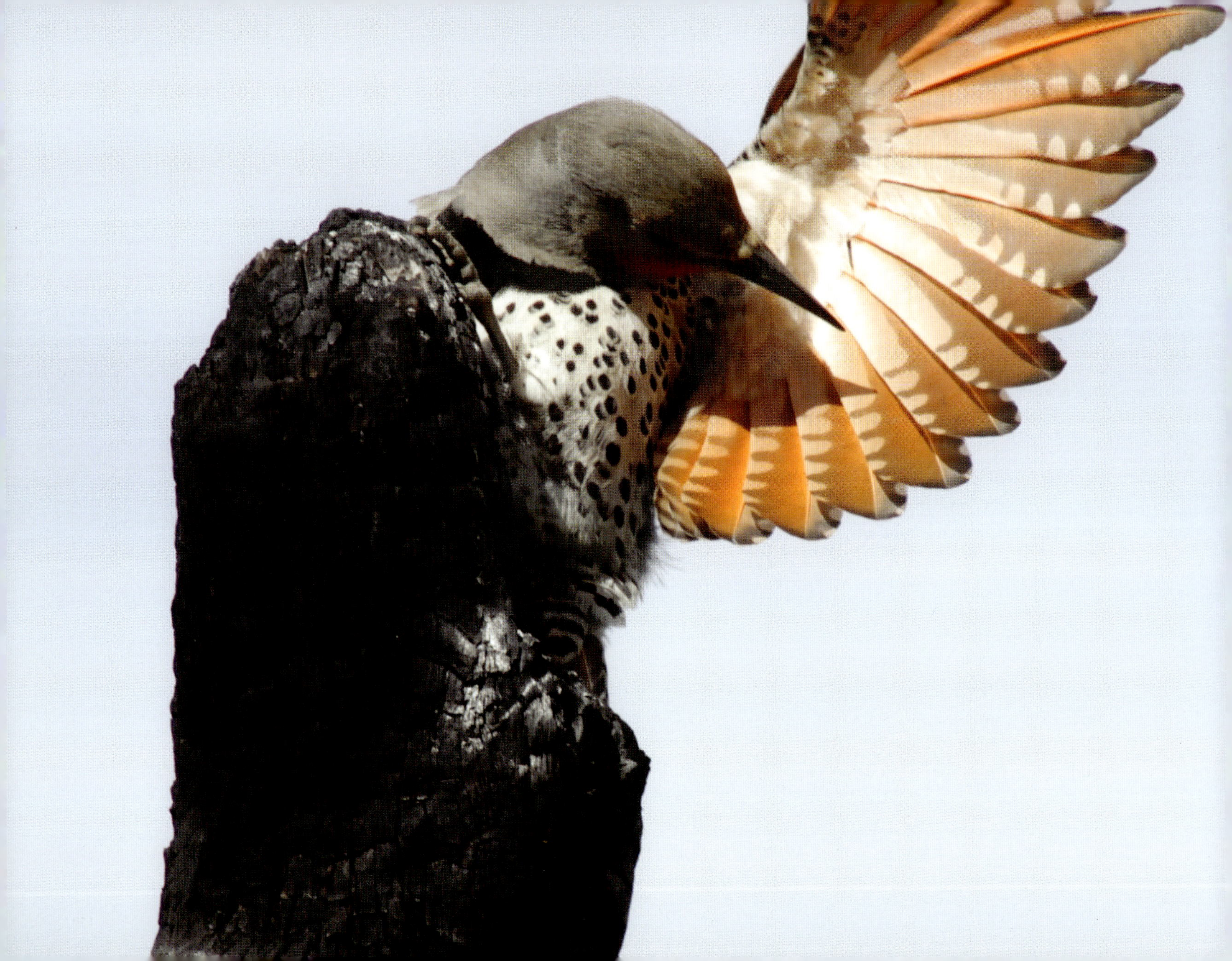

FURTHER READING

In ways that would surely please J Harlen Bretz (September 2, 1882–February 3, 1981), the ongoing scientific inquiries into the mysteries of the Pacific Northwest's ice age floods continue to fill in important gaps. Part of the challenge in writing *Beautiful Wounds* has been trying to avoid fostering the impression that it is a book about geology, rather than the story of a life-changing, photographic vision quest. That said, I'd like to broadly acknowledge and express my appreciation to the ranks of scholars and earth scientists whose work I've relied upon to build the geologic bedrock, so to speak, underlying this narrative and photography.

The contemporary bible of scabland geology is Bruce Bjornstad's *On the Trail of the Ice Age Floods*, which is a two volume work (2006 & 2012), the second with coauthor Eugene Kiver. *Cataclysms on the Columbia* (1986 & 1991) by John Eliot Allen, Marjorie Burns, and Sam C. Sargent is a well-crafted homage to Bretz that also offers a broad, regional perspective on the Bretz floods, especially how they left their marks on the lower Columbia River Gorge. John Soennichsen's *Bretz's Flood* (2008) is a first-rate biography of Bretz, and his 2012 publication, *Washington's Channeled Scablands Guide*, remains a valuable handbook. In addition to his numerous journal publications on the scablands, Victor Baker, Regents Professor of Geosciences at the University of Arizona, has been an invaluable scientific voice in popular documentaries, including *Mystery of the Megaflood*, a PBS *NOVA* episode about the Bretz story that first aired in 2005. Like thousands of others who follow his work

Left: Preening Northern Flicker.

online and on public television, I'm a devoted fan of Central Washington University's Senior Lecturer Nick Zentner, who has a rare gift for making earth science accessible and just downright fun. Special thanks to Catherine Cooper, Associate Professor of Geosciences at my alma mater, Washington State University, for reviewing the *Beautiful Wounds* manuscript prior to publication and offering helpful technical advice.

In the early 1990s, a consortium of civic and academic entities in Washington, Oregon, Idaho, and Montana created a regional network, the Ice Age Flood Institute, to nurture public interest in the remarkable story that Bretz delivered a century ago. The institute—which has grown to eight chapters across the four states—was largely responsible for Congressional authorization in 2009 of The Ice Age Floods National Geologic Trail that extends from ancient Lake Missoula in Montana to the Willamette Valley in Oregon.